**Asymptotic Realms
of Physics**

**Asymptotic Realms
of Physics**

edited by
Alan H. Guth,
Kerson Huang,
and
Robert L. Jaffe

**Essays in Honor of
Francis E. Low**

The MIT Press
Cambridge, Massachusetts
London, England

This book was set in Times Mathematics by Asco Trade Typesetting Ltd., Hong Kong and printed and bound by Halliday Lithograph in the United States of America.

Library of Congress Cataloging in Publication Data

Main entry under title:

Asymptotic realms of physics.

   "Research publications of Francis E. Low": p.

   1. Physics—Addresses, essays, lectures. 2. Low, Francis E. (Francis Eugene), 1921–   —
Addresses, essays, lectures. 3. Physicists—United States—Biography—Addresses, essays, lectures. I. Low, Francis E. (Francis Eugene), 1921–. II. Guth, Alan H. III. Huang, Kerson, 1928–. IV. Jaffe, Robert L.
QC71.A85   1983   530.1   82-13999
ISBN 0-262-07089-8

Contents

Preface   vii

An Appreciation   ix   Murray Gell-Mann

Francis E. Low—A Sixtieth
Birthday Tribute   xi   Marvin L. Goldberger

The Need to Know   xvii   Jeremy Bernstein

The Support of Basic
Research   xxv   Val L. Fitch

Research Publications of
Francis E. Low   xxix

**Why the Renormalization Group
Is a Good Thing**   1   Steven Weinberg

**The Physics of Asymptotic
Freedom**   20   Kenneth A. Johnson

**The Running Coupling
Constant in Quantum
Electrodynamics**   32   Sidney D. Drell

**Functional Formulation of the
Renormalization Group**   37   Kerson Huang

The Topological Bootstrap   49   Geoffrey F. Chew

**The Fixed Point of Classical
Dynamical Evolution and
Chaos**   70   Mitchell J. Feigenbaum

**The Thermal Conductivity of
Metals and Insulators**   86   Victor F. Weisskopf

A Simple Quantum-Mechanical
Problem   95   Herman Feshbach

How to Analyze Low-Energy   Robert L. Jaffe
Scattering   100

Compound Bags and Hadron-   Carleton DeTar
Hadron Interactions   118

The Chew-Low Theory and the   John F. Donoghue
Quark Model   128

The Simplicity of Bremsstrahlung   D. Danckaert,
Cross Sections   139   P. DeCausmaecker, R. Gastmans,
                                                   W. Troost, and Tai Tsun Wu

Gauge Invariance and Mass,   R. Jackiw
III   147

Gribov Ambiguities and   William I. Weisberger
Quantization in the Axial
Gauge   162

Beyond the Mystery of Quantum   Adrian Patrascioiu
Mechanics   178

The Simple Facts about the   Gino Segrè
Baryon Asymmetry of the
Universe   184

Speculations on the Origin of the   Alan H. Guth
Matter, Energy, and Entropy of
the Universe   199

Cosmological Consequences of   So-Young Pi
Grand Unified Theories   223

From Gell-Mann–Low to   Asim Yildiz
Unification   239

Preons and Supersymmetry   250   Jogesh C. Pati, Abdus Salam,
                                                   and J. Strathdee

Contributors   261

Preface

On October 16, 1981, the Massachusetts Institute of Technology held a symposium in honor of Francis E. Low on the occasion of his sixtieth birthday. The celebration also marked Francis's embarkation on a new career as Provost of MIT. Over three hundred friends and colleagues gathered that day to hear talks by Goeffrey Chew, Mitchell Feigenbaum, Val Fitch, Kenneth Johnson, Val Telegdi, and Steven Weinberg. The symposium was followed by a banquet at which Murph Goldberger described the early days of Francis's long and distinguished career.

This volume is a continuation of our celebration of Francis's life in physics. It contains essays by a few of his friends, colleagues, and students. Francis's contributions to theoretical physics have been both broad and deep. Many papers in this collection reflect his lifelong interests and pioneering accomplishments in physics: the renormalization group and the short-distance structure of field theories, the dynamics of hadronic interactions, the high-energy behavior of weak interactions. Other essays, inspired in one way or another by Francis, nevertheless range far from his own research interests. We value this diversity and hope that the volume contains some surprises for Francis as well as a record of his beneficent influence over the course of modern theoretical physics.

Alan H. Guth
Kerson Huang
Robert L. Jaffe

An Appreciation

It is a pleasure to add a few words to this volume that celebrates the life and work of Francis Low.

Few things have meant more to me than our friendship of thirty years' standing, the work we have done together, and the time we and our families have spent together.

Francis Low is remarkable for his unpretentious brilliance, his rare common sense, and his probing and skeptical mind. His achievements in science have been widely acclaimed, but not nearly so much as they deserve. No doubt they will be matched from now on by notable contributions to the development—or at least the preservation—of a singular institution for research and teaching, the one where I had just finished my graduate education when I first encountered Francis.

Not the least of the rewards of that meeting has been the opportunity to know and admire his wonderful wife, Natalie. I wish them both a splendid future.

Murray Gell-Mann

Francis E. Low—
A Sixtieth Birthday Tribute

Marvin L. Goldberger

About ten years ago, Francis, Murray Gell-Mann, and I were at a Coral Gables Theoretical Physics Conference. We were seated together listening to a talk by a great, but then aged physicist. It was awful. The three of us agreed that if any one of us was ever caught doing such a thing, the others had the obligation to shoot the speaker through the head. The reason that Francis and I are now administrators who are never asked to give physics talks is that we are both sure that Murray really would shoot us.

What I am going to tell you about Francis here is a highly personal version of the past nearly thirty years. I don't want to leave you with the impression that portions of his life in which I was not involved are not interesting and conceivably important, unlikely though that may be. But, in the interest of your sanity and patience, I shall stick to what I personally know to be true.

I first met Francis in 1952 when he passed through Chicago on his way to Urbana where he was being looked over for an appointment. In retrospect, I don't really know why Murray and I didn't try to recruit him for Chicago. Murray, of course, was still a child, but I should have been more alert. I remember Rabi was talking three years earlier about the fact that they had put their best student at Columbia (who was Francis) working on some dumb problem suggested by Aage Bohr (which Francis solved), and I knew that work and some other things he had done. But Chicago's loss was clearly Illinois's gain. Some of Francis's most important work was done at Illinois by him alone, and with various collaborators, especially Geoffrey Chew who had gone there a year earlier.

My first serious contact with Francis was via extrasensory perception in the summer of 1954. I had been at Brookhaven early in the summer working the problem of the low-energy limit of Compton scattering from spin one-half particles. That spring Murray Gell-Mann and I had made the conjecture that through first order in the frequency the scattering amplitude could be completely described in terms of the particle's charge, mass, and anomalous magnetic moment. We had not completed the proof when Murray ran off to practice his spoken Erse off the Scottish shores, and I finally finished it at Brookhaven. Francis then came and sat at the very same desk and solved the same problem. A suspicious person who didn't believe in ESP might think he was slinking around waste baskets and came upon my discarded notes. The answer to that argument is that he solved the problem in a much deeper and more profound way, completely superior to the clumsy (though brilliant) methods Murray and I had used. I must confess I've never asked Francis how he came to be working on this problem in the first place—Murray and I came upon it in connection with our work on dispersion relations.

The first nonspiritual association Francis and I had was in the summer of 1955. Ken Watson recruited Keith Brueckner, Geoff Chew, Francis, and me to work on what was then the classified program on controlled fusion. It was a rather bizarre experience. Ken had spent the previous term at Los Alamos and was our undisputed leader, and somehow Keith had become an expert on plasma physics. Geoff, Francis, and I didn't know beans about the subject. We, however, approached it with our characteristic modesty and quickly concluded that the practitioners didn't really know what they were talking about. We had a particularly brutal encounter with Ed Frieman, then at Princeton's Project Matterhorn, which after several days of wrangling left the three of us with our tails between our legs. I remember one morning after our comeuppance, sitting around and hearing Geoff say, "Let's get that little bastard down here and see how much he knows about mesons!"

We did, however, lick our wounds and ultimately wrote a brilliant paper on how to deduce the equations of magnetohydrodynamics from the Boltzmann equation. I can only say that it has become a classic. I think, in fact, we probably had more reprint requests for that work than anything the three of us were ever involved in before or after.

There are two interesting sidelights on the work we did on plasma physics. First, this famous paper was published in the *Proceedings of the Royal Society*, I think the only time any of us has appeared in that distinguished journal. The paper was submitted by Chandrasekhar, who is, of course, a member. Furthermore he, Chandra, insisted on writing in the equations, thinking, probably correctly, that we weren't competent to abide by their rather arcane rules. The second part of the episode is shrouded in mystery. We wrote, about two years later, another paper which was really a sequel to the first and, we thought, equally brilliant. Again, Chandra was the sponsor and the filler-in of equations. The paper has never appeared in print. None of us ever had the nerve to ask Chandra what happened. Did he lose it? Did he mess up the equations so badly that the paper was rejected for rule violation? Was it suppressed by an enraged referee who resented particle physicists mucking around his turf? I don't really know where the truth lies.

The summer was not only a scientific success, but it was our first taste of western living with hiking, jeep riding, picnics, etc. We all had small children, and we became a kind of extended family. It was where Mildred and I first got to know Natalie, when she was just an Ms., and not yet a Doctor. She was always the baby of the group, much younger than the rest of us, but very precocious. It's still true.

During the two years following the Los Alamos summer, Francis, Geoff, and I, together with Yoichiro Nambu, attacked the problems of pion-nucleon scattering and photomeson production using the techniques of dispersion theory. The work culminated in the publication of two papers in 1957 referred to by the totally unpronounceable acronym CGLN. At about the same time Francis was called to MIT, as they say in the East, and I was called to Princeton.

We continued to share summers with the Lows, sometimes in connection with the Jason group in Maine, Woods Hole, and elsewhere. But our next, and in some ways our most intense interaction came during 1962–63. Mildred was invited to become a Radcliffe Scholar, an innovative program initiated by Mary Bunting—called Polly, naturally, by those who knew—designed to rescue and reinvigorate fallen women, those who had deserted lives of creativity for procreative activities or what have you. It was my first experience as a spouse, and I was doubly fortunate that Princeton would give me a leave of absence and that MIT graciously took me as a Visiting Professor.

The previous summer, in Berkeley, Murray and I observed that the conjectured Regge-pole behavior of scattering amplitudes at high energies might be understandable in the framework of quantum field theory. Furthermore, we conjectured that radiative corrections involving massive vector particles could cause the fixed singularities in the angular-momentum plane, ordinarily associated with the field-theoretic "elementary" particles, to in fact become moving singularities or Regge poles. Francis and I began a more elaborate and systematic study at MIT, while Murray and Fred Zachariasen were carrying one on at Caltech. We were in constant communication and were later joined in the work by a Caltech student whose name was Egon Marx, always referred to by us as Karl. Murray came to MIT during the second term of the year, and we really tackled the problem in earnest. We were having a ghastly time with the scattering of massive vector mesons by fermions. Superficially, the various amplitudes had wild high-energy behavior, all the terms of which canceled leaving the logarithmic behavior we were seeking. We were inundated with hideous arithmetic, however, and would never have gotten the right answer had Francis not had a brilliant inspiration for a choice of gauge; we then solved the whole problem one Sunday afternoon and evening.

This resulted in another unsolved publication mystery. We wrote an extremely cryptic *Physical Review* letter. It was rejected by a referee as being unintelligible. We knew that that was unfair, because we had sent a copy to Stanley Mandelstam, and he understood it. We ultimately submitted it to *Physics Letters*, and they were proud to have it. I have never stopped, even after 18 years, trying to find out who that ignorant, but admittedly rather gutsy, bastard of a referee was. Murray vowed never to publish in *Physical Review Letters* again, but I'm not sure he kept his pledge.

A high point of that year was a gourmet dinner at the house where the Gell-Manns were living (which was Schwinger's house). Everyone came in formal attire, only champagne was served before dinner, and wine so fine, supplied by Irwin, better known as E. Phillips, Oppenheim, that he would not allow it to be transported in Murray's miserable car for fear of breaking the long molecules.

The scientific and social lives of the Lows and Goldbergers crossed again in 1967–68 when Francis was a visitor at the Institute for Advanced Study. We had a marvelous time socially, while scientifically we failed to solve the problem of the divergence difficulties encountered in weak interactions. We were joined in that work by Norman Kroll and Murray, and although great cleverness was shown, we never came close to what we now regard as the final solution of that problem the red-haired fellow found. Francis and I did solve a longstanding problem related to the high-energy limit of forward Compton scattering from a bound electron.

During that year at Princeton, Francis fulfilled a lifelong ambition of his, learning to fly. As a result of an old skiing injury, he was washed out of flight training during the war. This, incidentally, caused him to serve as a combat soldier in Italy. Having successfully mastered flying, he took every opporunity to use his plane to go to meetings. On one occasion while he, Geoff Chew, and I were collaborating on multiperipheral Regge exchange models and were scheduled to meet in Boston, I was presented with a terrible dilemma. I was to meet Francis at a conference at the University of Massachusetts in Amherst and fly back with him to Boston. He had had his license for about two days at the time, and although my general confidence in Francis's ability to do anything he puts his mind to was virtually unlimited, I was more than a little apprehensive about his lack of experience. A mutual friend and longtime pilot, Hal Lewis, had told me never to fly with anyone who had less than 1,000 hours of flying time. I was saved, however, by the fact that the weather was so bad that Francis couldn't fly to the conference, and we all had to drive to Cambridge.

I could go on longer, talking about incidents where our world lines have crossed. Instead, I'll close by simply reminding you of what a multifaceted, talented man we are honoring here tonight. He is a superb pianist, skier, tennis player, songwriter, linguist, and pilot. He is also quite good as a physicist. I have often envied these many abilities of Francis and attributed my own inadequacies in several of the areas to the fact that he is so much older than I am, and has been at them longer.

Quite seriously, Francis is an unusual person, and I feel honored to have him as a friend and colleague. Finally, I promised Natalie that I would mention her frequently. I know how important a role she has played in Francis's life and in the life of Mildred and me. She is also as excellent a charades player as Francis is—and I forgot to mention this last and greatest talent. Having done so, I can at long last stop and propose a toast to the continued health and success of our friend.

The Need to Know   Jeremy Bernstein

Well before I actually met Francis Low he was already a hero of mine. The reason for this may seem odd to anyone who did not grow up, as I did, in the theoretical physics of the 1950s. Francis had actually gone *beyond* perturbation theory! One must remember that this was a time when an entire industry of theses—including my own—conference talks, and even books was based on the proposition that expanding things in a series in the meson-nucleon coupling constant—a number that was taken to be at least 10—might lead to something interesting. In retrospect, most of this work resembles the sort of craniometry that was fashionable in the nineteenth century and made about as much sense. But there are very few Ph.D. theses which, in retrospect, make a lot of sense. I recall remarking to my roommate Chuck Zemach that I was going to call my thesis Deuteronomy since it dealt with the deuteron, to which Zemach responded that he was planning to call *his* Exodus. In any event the Chew-Low theory changed all of that, and Francis became a hero of mine. I also have an eerie recollection of Geoffrey Chew coming to Harvard to talk about the theory and hearing one of the senior faculty asking, after the lecture, if Low was his graduate student.

The first time I actually met Francis was, I believe, in the spring of 1957. He was at MIT for a visit just before joining the department. It may come as a surprise, considering the megalithic enterprise that particle physics has now become in Cambridge, how few we were in those days. At Harvard there was, of course, Julian Schwinger and a platoon of graduate students. There were Roy Glauber, Bob Karplus, and Abe Klein as junior faculty and, as even more junior faculty, Ken Johnson and, later, Paul Martin. I was ensconced at the Cyclotron Laboratory as the house theorist, an absolutely wonderful job with no responsibilities whatsoever. On the faculty at MIT there was, of course, Vicki Weisskopf, Herman Feshbach, Felix Villars, Sid Drell, Bernie Feld, and perhaps a few others. There were just about enough people at the two institutions to fit around one of the large tables at *Chez Dreyfus*, where several of us would meet each Wednesday to have lunch. During these lunches Julian would fill us in on what he had been doing, with Vicki translating for the rest of us. Both parties wrote furiously on the paper placemats, which I often made off with to study later. I still have a few. To this amalgam Francis was added. Indeed, I have a distinct recollection of his first appearance at the table. Julian was lecturing, and I was straining every neuron to follow the gist. Francis was sitting next to me, and all of a sudden he began to say to me—I think it was to me—"The man [Julian] is wrong. What he is saying is wrong." Julian was also a hero of mine, which left me in a state of psychological ambiguity. I have no recollection of what the issue was, but I recall that it was amicably resolved and that our lunches were never quite the same.

That spring was my last in Cambridge, and before moving on to the Institute for Advanced Study, I had the summer free, and I arranged to spend the time as an "intern" at Los Alamos. There was no real way that I could be fitted into any of the existing programs in the lab, so I was assigned an office in the Theoretical Physics Division, which I shared with Ken Johnson. In the course of things Francis suggested that Ken and I work on a problem in π^0 decay that we spent the summer working on.

I should try to describe the rather odd emotions I had then about nuclear weapons. To understand them one must have a feeling about how being at Los Alamos seemed to a young physicist. Los Alamos was then still very much a closed community. There were fences within fences. Much of the "temporary" construction used during the war was still standing, including the barrackslike housing project in which I lived. All of us who worked there had "Q-clearances," which were the highest clearances then given out by the Atomic Energy Commission. This meant—and it had been the policy of the laboratory from the days of its first director, Robert Oppenheimer—that we had access to classified information on a "need-to-know" basis. In the work that I was doing Johnson there was nothing classified that we needed to know, so I didn't learn any secrets. Low was working on fusion, and he and the other senior people were at that time fairly optimistic that it could be made to work for power production. Access to the fusion program was restricted, too, but Ken and I were allowed to attend some of the fusion seminars. For a young physicist, as I was, going to them gave me the feeling of being admitted to some sort of secret brotherhood—the brotherhood of people who needed to know. I don't recall any discussions of atomic weapons, but one had a sense of their presence. Indeed, I think some were stored near the lab in a cave surrounded by barbed wire and guarded by machine guns. As strange as this may now appear, the weapons represented a brooding and almost romantic presence. I, at least, lost sight of what they were and what they were for. They became symbols of the brotherhood of the need to know. Being part of the brotherhood gave me a somewhat superior feeling toward people—most people—who did not need to know. I do not think that this attitude was characteristic of most of the senior people, who had any real responsibility for the larger decisions, but I am trying to describe how things then seemed to me—the seduction of power.

In any case, when Francis announced one day that he was going to Nevada to see some atomic bomb tests, I decided that I would very much like to go along. There was absolutely no reason for my being there except curiosity. But this was also his reason for going. He promised that he would speak to Carson Mark, then the director of our division. He has now retired. In due time, I was told that I could go if I agreed to pay my expenses—the commercial flights and so on—since there was no reason to charge them to the laboratory. Thus it was that sometime in August, Francis and I and Carson Mark left Los Alamos in a light plane that took us to Albuquerque, where we took a commercial flight to Las Vegas.

By the time we reached Las Vegas, it was about nine at night. A test had been scheduled for the following morning, but it had become customary for the people who were working at the test site to spend their free time in Las Vegas. Since the tests were scheduled just before dawn to insure the right light for photography, and since there was nothing for us to do at the site anyway, we joined some of the other people from Los Alamos to play blackjack in the casinos. Not long before my summer at Los Alamos it had been shown, mathematically, that blackjack is the only casino gambling game in which a successful strategy can actually be made. It is fairly complicated, and makes use of the fact that in casino blackjack the dealer is, in effect, an automaton who follows preset rules. In any event, the theoreticians at Los Alamos had developed the strategy empirically by playing thousands and thousands of hands on one of the computers in its off hours. (The same strategy had also been worked out independently by pure mathematical methods.) This had resulted in the preparation of a summary strategy card of which people going to the test site could get a copy. If one followed the rules one would be assured, at least, of losing one's money at the slowest possible rate. I have never enjoyed card games very much, but this one appealed to me as a mathematical exercise. I do not have any recollection of having won anything Francis remembers that he won about ten dollars. About 1:00 A.M., there was some prearranged signal in the casino, and we all left by automobile for the test site.

The site was located in Mercury, Nevada, about two hours by car from Las Vegas. We arrived there about 3:00 A.M. and went at once to some sort of central control building. There was a large room which had cots, and I remember talking to a meteorologist about the weather for the next morning and then dozing off for a couple of hours. Someone woke me up—perhaps it was Francis—and, still in darkness, we went outside. We had been told that all of the Los Alamos bombs that summer had been named after famous scientists. (The bombs from the Livermore Laboratories in California had been named after mountains.) In fact, Carson Mark had told us that the following afternoon we would be able to go to a tower and visit "Galileo"—a bomb that was to be exploded later in the series. I suppose that I regarded the prospective explosion as some sort of fireworks display. Viewed now, from a perspective of twenty years, I find it difficult to understand how I felt then, but I am sure that I had not thought of bombs as weapons designed to kill people. I once had a long talk about atomic weapons with Stanley Kubrick after he had made *Doctor Strangelove*. He said that, in his view, most of the time we make a successful psychological denial of their existence. They become abstractions. They become something that we cannot think about, because we don't want to think about them. I had read a great deal about nuclear weapons and about Hiroshima, but I hadn't experienced anything. It was all words and pictures.

It was quite cold that morning. Carson Mark told us to face away from the blast and then, after it came, to count to ten. To look at the explosion before that would be to risk blindness. I think that we had some sort of smoked glass, and after ten we could turn around and look through the glass. There was a loudspeaker that reported on the time left before the blast: "T-minus ten minutes"—something like that. The last few seconds were counted off one by one. We had all turned away. At zero there was the flash. I counted and then turned around. The first thing I saw was a yellow-orange fireball that kept getting larger. As it grew, it turned more orange and then red. A mushroom-shaped cloud of glowing magenta began to rise over the desert where the explosion had been. My first thought was, "My God that is beautiful!" I wonder if any of the citizens of Pompeii said that to themselves in that first unguarded moment after the volcano erupted. It then struck me that there had been no sound. We were some ten miles away, and the light had gotten to us first. While I was thinking this, someone said "The shock wave!" and the next thing I remember was a rather painful click in my ears as the shock wave from the explosion passed by us. I do not have any recollection of any real sound. I didn't notice any—just this sudden pressure in my ears. By this time, the sun had risen almost as if it had been summoned by the explosion. The cloud had now turned into a leaden purple-black—a malignant mass of radioactive debris, pure and undiluted death. After looking at it for some time, I went inside.

I wanted to be alone with my thoughts and went back to get some sleep in the dormitory room I had been assigned. I had not really been to sleep for something like twenty-four hours, so when I woke up it was early afternoon. Francis and I went to find Carson Mark, and he proposed that we take a tour of the test site. I have not retained a very clear impression of the test site as a whole. I remember that it was set in the scrub desert. There were a few hills around covered with yucca trees. Some tunnels had been dug into some of the hills for underground tests. Both Francis and I remember the light. It was exceedingly bright and flat. I do not recall having seen any animals. There were a great many military people around. This was a time when soldiers were being used in maneuvers near the test explosions. People were in their shirt sleeves.

I have retained a very clear memory of "Galileo." The bomb was located on top of a metal tower perhaps a hundred feet high, about ten miles from the buildings. I had not expected to be allowed to see an unexploded bomb, and I made a remark to Francis about "visiting the old gentleman in his tower" that he still recalls. We drove to the tower by car. I recall that there was a crude elevator that took us most of the way up the tower to a platform. I also remember that Francis and I climbed up some kind of metal ladder suspended in midair to get to the top of the tower. On that ladder I got a sudden feeling of acrophobia, of near panic that contributed to a growing feeling of uneasiness as we approached the bomb. "Galileo" looked like a slightly oversized cannister vacuum cleaner, with wires and cables leading from it. I also remember that there were pumps—perhaps vacuum pumps or a cooling system—that made rhythmic clicking sounds. There were a few men in the tower making adjustments of some sort on the bomb. I recall being told that both "Galileo" and the bomb we had seen explode in the morning were substantially more powerful than the two fission bombs that had actually been used in World War II. It was difficult to make any connection between this machinery we were looking at and a weapon of war. I had by this time been around complicated scientific equipment for several years and, on the scale of a cyclotron, for instance, the bomb did not look very impressive. After a while, we had seen what there was to see of "Galileo" and continued on with our tour.

As we drove around the site, I noticed that it had been divided up by zones with signs indicating levels of radioactivity, presumably corresponding to how long it was since the area had been used for an explosion. There was no way to tell by looking which parts of the desert were relatively safe and which were not. While the areas had been marked off with signs, other than that there was no visible difference between one sere burned-off patch of desert and the next. Of the tower that had held the bomb we had seen exploded in the morning there was not a trace. It had been vaporized.

Then came an experience about which I and Francis remember different details. We do not necessarily disagree, but they have arranged themselves differently in our minds. It is so vivid to me that as I write this I can see the scene perfectly. I am sure that it happened as I remember it. Carson Mark took us to a low, concrete building somewhat separated from the rest. This was the building where the atomic bombs were assembled and stored. When we walked inside, I remember looking and recoiling toward the door. I also remember Francis telling me that if they exploded, being a little farther away was not going to make much difference. I have no idea how many of them there were: very likely enough to devastate a country. The center of a bomb is a perfect sphere of uranium or plutonium. I believe it was called the "pit." The pits looked like shiny bowling balls. Around the pit was to be wrapped a high-explosive shell, carefully shaped. In the building, we saw—and I am sure of this—a man filing on the high explosive that had been wrapped around a pit. Francis remembers him as whittling on the explosives. Next to him there was a woman knitting. My thought at the time was that maybe she was his wife and that they had agreed she would be with him while he worked so that if it all went up they would go together. When I wrote recently and asked Carson Mark if he recalled this scene, he replied that he did not but he was quite sure that it could not have been a man and his wife since he did not remember any such couple in the testing activity. He thought that it might have been a secretary visiting from another area. Francis does not remember the woman at all. But she was there. I also remember that I was given a pit to hold briefly. Although it was heavy,

I was able to hold it. It was slightly warm to the touch. There was something so strange, both human and so inhuman, about this scene—I thought, is this the way the world ends?—that it made more of an impression on me than either of the explosions we saw. The next morning, a bomb named after a mountain—I think "Shasta"—was suspended in a balloon; when it exploded, it lit up and consumed an entire hillside of yucca trees.

There is an epilogue to all of this. In 1963, the United States and the Soviet Union signed a limited test-ban treaty. The signing of this treaty followed five years of debate, both within our government and with the Soviet Union. Now it seems all but inconceivable that such a debate was ever necessary. But at the time I believed, and my attitude was not uncommon, that these above-ground tests were somehow essential for our national security, given the state of the Cold War. Therefore, when Adlai Stevenson, for whom I had voted in 1952, took the position in his election campaign of 1956 against Eisenhower that above-ground testing should be stopped, I thought that on this issue he was wrong. I simply had not understood the long-range implications of the open-air testing.

As it happened, at the end of that summer of 1957 I met Stevenson. I was then going out with a girl who lived near Chicago, and since Francis was driving back to Boston I hitched a ride with him as far as Chicago and went to visit my friend, who was at home with her family. They knew Stevenson well, since he lived nearby. After listening to my description of the test, they thought that Governor Stevenson would be also interested and they took me to a party to meet him. After we were introduced, I began to tell him about the tests and the need for the tests. He listened for about five minutes and then simply got up and walked away without saying a word. I have never forgotten that. At the time I was enormously embarrassed, and when I think of it now I am embarrassed still. But I now think I understand why he walked away. He must have seen in me a young man made foolish by his proximity to absolute power—the absolute power of an atomic bomb. He was right. Proximity to absolute power can make fools of any of us.

Postscript

When I was trying to reconstruct these events, which took place nearly twenty-five years ago, there was something that I now find that I left out. I did recall vaguely that there had been some sort of military presence in the desert, but I could not remember its purpose, if ever I knew it. Recently

I came across a book—*Atomic Soldiers* by Howard L. Rosenberg (Beacon Press, 1980)—which explains what Francis and I witnessed at Los Alamos. The explosion on August 31 of "Smokey"—a forty-two-kiloton bomb— more than the combined power of the two bombs that fell on Hiroshima and Nagasaki, and the explosion of the eleven-kiloton "Galileo" on September 2, were the last above-ground tests in the United States that were accompanied by troop maneuvers. Indeed, about three thousand troops were stationed within three miles of the places where these bombs were exploded, and soon after, the soldiers were moved to within a few hundred yards of ground zero. This was done, apparently, in an attempt to determine how soldiers would perform, psychologically and otherwise, under the conditions of atomic warfare. It is difficult to understand in retrospect how a maneuver like this could ever have made sense to any- body. At the present time, several of these former soldiers, some of whom are dying, are attempting to sue the government. As I write, the disposi- tion of these suits remains undecided.

The Support of Basic Research   Val L. Fitch

It is a very special pleasure to participate in these festivities honoring Francis Low. Like most of his peers I have been educated by Francis, and it is highly appropriate that he describes himself in *Who's Who* as an educator. As an example I can point to a book Francis authored about 15 years ago entitled *Symmetries and Elementary Particles*, published by Gordon and Breach. In this little book he discusses in considerable technical detail the discrete symmetries: charge conjugation C, space inversion or parity P, time reversal T, and, of course, TCP. In view of my principal interests during these years that book has been exceedingly useful.

But here I would like to address a topic that is of great concern to Francis in his new role as science administrator: the state of funding for basic research.

The current problems associated with funding basic research are not new. I recently had the occasion to learn something about Joseph Henry, the discoverer of self-induction. In his day, Henry was the closest of any American to being a pure scientist. The emphasis in the United States had been on useful knowledge, applied science and, except for isolated examples such as Franklin and Jefferson, there had been very little interest in the fundamentals. Tocqueville, in his *Democracy in America*, in 1835, has a chapter entitled, "Why Americans Prefer the Practice Rather Than the Theory of Science." He went on to observe there were very few calm spots in America for reflection and meditation, factors necessary for pursuing pure science. In 1846, at the age of 49, Henry left Princeton and became the first Director of the Smithsonian Institution. It was a grant from the Englishman Smithson of £100,000 in 1835 that established this first organization in the United States to be devoted to science. Smithson specified that it be "for the increase and diffusion of knowledge among men." Henry followed these instructions closely. He said, "There's another division with regard to knowledge which Smithson does not embrace in his design, namely the application of knowledge to useful purposes. He has found it not necessary to found an institution for this purpose. As he said, there are already in every civilized country establishments and patent laws for the encouragement of this kind of mental industry. As soon as any branch of science can be brought to bear on the necessities, conveniences, and luxuries of life, it meets with encouragement and reward. Not so with the incipient principles of science."

Obviously I have seized on these observations because Henry is making a strong pitch for basic research. The pressures in Congress are today the same as they were in Henry's day. Basic research having no political base, no constituency, has always had tough sledding in Congress. Smithson's bequest was made in 1835. It took 10 years of wrangling before Congress finally accepted and then set up the Smithsonian Institution. Originally it was charged by Congress to promote everything, but Henry managed to see that once something was organized the activity was shed. Thus, such institutions as the National Gallery of Art, the Corcoran Gallery, the Library of Congress, the Weather Bureau, and the Bureau of Standards were spawned at the Smithsonian and then severed for lives of their own.

Today the National Science Foundation has replaced the Smithsonian as the premier national organization for promoting basic research. And most recently it has been busy responding to pressures from Congress to support more applied research. Tocqueville's observation in 1835 still holds. The pressures experienced by Henry still exist unabated.

Flying in the face of Henry's admonitions, the National Science Foundation currently supports, as only one example, geotechnical engineering at a time when the oil companies are spending enormous sums in similar areas. Yet fundamental basic research activities go begging for support. Physics experiments are the great voyages of discovery of our day, but at present physics draws less than 7 percent of the NSF budget.

There is another criticism that I would like to level at the National Science Foundation. Whether one likes it or not, pure science is an elitist activity, almost by definition. It is an activity that is done much better by some people than by others. But the pressures on the National Science Foundation are always to make it more egalitarian, for example to distribute the funds by geography rather than by talent. This, too, is not a new phenomenon. Henry Roland, who succeeded Henry as leading scientist in the United States in the last century, in testifying before Congress said, "What do you want? A flight of eagles or a swarm of mosquitos?"

Suzanne Garment, a writer for the *Wall Street Journal*, recently devoted a column to these concerns for basic research. She had just visited the radio telescope at Arecibo in Puerto Rico. She wrote:

Here I was, fresh from a city [Washington] composed almost wholly of memos, intrigues, and lunch, looking into the face of the startling stone and metal beauty these scientists had created to stand against the surrounding wildness. The sight was more than visually striking; it was also the most profoundly embarrassing reminder of the craft, the self-discipline and the decades-long patience that create and maintain such an enterprise.

These are not qualities to be found in abundance these days in our society as a whole, much less in the Washington that purports to govern it. Yet it is on the cultivation of such traits, no less than on the specific results of scientific projects like this one, that the success of the Western enterprise has been carried.

... few people (in Washington) even know the words to describe and distinguish among long-term scientific and technological endeavors, let alone treat them as a matter of national concern.

Much of the public indifference to basic research can be blamed, perhaps, on scientists themselves. Certainly what we consider fundamental research cannot be differentiated by the public. And this is a matter of education in its broadest sense. So we are fortunate, indeed, to have Francis in this new position as administrator and spokesman for science.

Research Publications of Francis E. Low

On the Ground State of The Deuteron, *Phys. Rev.* 74: 1885–1886 (1948)

On The Effects of Internal Nuclear Motion on The Hyperfine Structure of Deuterium, *Phys. Rev.* 77: 361–370 (1950)

Singular Potentials and the Theory of The Effective Range (with K. Brueckner), *Phys. Rev.* 83: 461–2 (1951)

On The Hyperfine Structure of Hydrogen and Deuterium (with E. E. Salpeter), *Phys. Rev.* 83: 478 (1951)

Bound States in Quantum Field Theory (with M. Gell-Mann), *Phys. Rev.* 84: 350–354 (1951)

Natural Line Shape, *Phys. Rev.* 88: 53–57 (1952)

Motion of Slow Electrons in a Polar Crystal (with T. D. Lee and D. Pines), *Phys. Rev.* 90: 297–302 (1953)

Mobility of Slow Electrons In Polar Crystals (with D. Pines), *Phys. Rev.* 91: 193–194 (1953)

Bremsstrahlung At High Energies (with H. A. Bethe and L. Maximon), *Phys. Rev.* 91: 417–418 (1953)

Quantum Electrodynamics at Small Distances (with M. Gell-Mann), *Phys. Rev.* 95: 1300–1312 (1954)

Scattering of Light of Very Low Frequency by Particles of Spin $\frac{1}{2}$, *Phys. Rev.* 96: 1428–1432 (1954)

Boson-fermion Scattering in the Heisenberg Representation, *Phys. Rev.* 97: 1392–1398 (1955)

Spatial Extension of the Proton Magnetic Moment from the Hyperfine Structure of Hydrogen (with W. M. Moellering, A. C. Zemach and A. Klein), *Phys. Rev.* 100: 441–442 (1955)

The Boltzmann Equation and the One-Fluid Hydromagnetic Equation in the Absence of Particle Collisions (with G. F. Chew and M. L. Goldberger), *Proc, Roy. Soc. A.* 236: 112–118 (1956)

Effective Range Approach to the Low-Energy *p*-wave Pion Nucleon Interaction (with G. F. Chew), *Phys. Rev.* 101: 1570–1579 (1956)

Theory of Photomeson Production at Low Energies (with G. F. Chew), *Phys. Rev.* 101: 1579–1587 (1956)

Application of Dispersion Relations to Low-energy Meson-Nucleon Scattering (with G. F. Chew, M. L. Goldberger and Y. Nambu), *Phys. Rev.* 106: 1337–1344 (1957)

Relativistic Dispersion Relation Approach to Photomeson Production (with G. F. Chew, M. L. Goldberger and Y. Nambu), *Phys. Rev.* 106: 1345–1355 (1957)

Decay of π Meson and Universal Fermi Interaction (with K. Huang), *Phys. Rev.* 109: 1400–1402 (1958)

A Lagrangian Formulation of the Boltzmann-Vlasov Equation for Plasmas, *Proc. Roy. Soc. A.* 248: 282–287 (1958)

Bremsstrahlung of Very Low-Energy Quanta in Elementary Particle Collisions, *Phys. Rev.* 110: 974–977 (1958)

Unstable Particles as Targets in Scattering Experiments (with G. F. Chew), *Phys. Rev.* 113: 1640–1648 (1959)

Proposal for Measuring π^0 Lifetime by π^0 Production in Electron-Electron or Electron-Positron Collisions, *Phys. Rev.* 120: 582–583 (1960)

Correspondence Principle Approach to Radiation Theory, *Am. J. Phys.* 29: 298–299 (1961)

Persistence of Stability in Lagrangian Systems, *Phys. of Fluids* 4: 842–846 (1961)

Limit on High-Energy Cross Section from Analyticity in Lehman Ellipses (with O. W. Greenberg), *Phys. Rev.* 124: 2047–2048 (1961)

Bound States and Elementary Particles, *Nuovo Cimento* 24: 678–684 (1962)

Comments on Chew's Bootstrap Relationship, *Phys. Rev. Lett.* 9: 277–279 (1962)

Elementary Particles of Conventional Field Theory as Regge Poles II (with M. Gell-Mann, M. L. Goldberger, and F. Zachariasen), *Phys. Lett.* 4: 265–267 (1963)

Elementary Particles of Conventional Field Theory as Regge Poles III (with M. Gell-Mann, M. L. Goldberger, F. Zachariasen, and E. Marx), *Phys. Rev.* 133: B145 (1964)

Elementary Particles of Conventional Field Theory as Regge Poles IV (with M. Gell-Mann, M. L. Goldberger, F. Zachariasen, and V. Singh), *Phys. Rev.* 133: B161 (1964)

Renormalization Group Method of High-Energy Scattering, *Zh. ETF* 3: 25 (1964). English translation in *JETP* (*Soc. Phys.*) 19: 579 (1964)

A Boson Selection Rule (with J. B. Bronzan), *Phys. Rev. Lett.* 12, 522 (Spring, 1964)

The Vacuum Trajectory in Conventional Field Theory (with M. Gell-Mann and M. L. Goldberger), *Rev. Mod. Phys.* 36: 640 (1964)

Exact Bootstraps In Some Static Models (with K. Huang), *Phys. Rev. Lett.* 13: 596 (1964)

Exact Bootstrap Solutions in Some Static Models of Meson-Baryon Scattering (with K. Huang), *J. Math. Phys.* 6: 795–816 (1965)

Heavy Electrons and Muons, *Phys. Rev. Lett.* 14: 238 (1965)

Are Wave Functions Finite?, in *Preludes in Theoretical Physics: In Honor of V. F. Weisskopf*, North-Holland, Amsterdam (1965)

Limits of Quantum Electrodynamics, in *Perspectives in Modern Physics* (Essays in Honor of Hans A. Bethe), edited by R. E. Marshak, Wiley Interscience, New York (1966)

Higher Symmetries, *Acta Physica Austriaca*, Supplementum III (1966)

Current Algebras in a Simple Model (with K. Johnson), *Prog. Theor. Phys.*, Suppl. 37 and 38 (1966)

Weinstein's Predecay Mixing Effect (with G. Jona-Lasinio), *Phys. Rev.* 152: 1411 (1966)

Current Algebra and Non-Regge Behavior of Weak Amplitudes (with J. B. Bronzan, I. S. Gerstein, and B. W. Lee), *Phys. Rev. Lett.* 18: 32 (1967)

Electromagnetic Mass Difference of Pions (with T. Das, G. S. Guralnik, V. S. Mathur, and J. E. Young), *Phys. Rev. Lett.* 18: 759 (1967)

Current Algebra and Non-Regge Behavior of Weak Amplitudes II, (with J. B. Bronzan, I. S. Gerstein, and B. W. Lee), *Phys. Rev.* 157: 1448 (1967)

Radiative Corrections to β Decay and the Quantum Numbers of Fields

Underlying Current Algebra (with K. Johnson and H. Suura), *Phys. Rev. Lett.* 18: 1224 (1967)

High-Energy Limit of Photon Scattering on Hadrons (with H. D. I. Abarbanel, I. J. Muzinich, S. Nussinov, and J. H. Schwarz), *Phys. Rev.* 160: 1329 (1967)

Helicity Poles, Triple-Regge Behavior and Single-Particle Spectra in High-Energy Collisions (with C. E. DeTar, C. E. Jones, J. H. Weis, J. E. Young, and Chung-I Tan), *Phys. Rev. Lett.* 26: 675 (1971)

Two-Particle Distributions and the Nature of the Pomeranchuk Singularity (with D. Z. Freedman, C. E. Jones and J. E. Young), *Phys. Rev. Lett.* 26: 1197 (1971)

Generalized $O(2, 1)$ Expansion for Asymptotically Growing Amplitudes (with C. Edward Jones and J. E. Young), *Ann. Phys.* 63: 476 (1971)

Analytic Continuation in Helicity and $O(2, 1)$ Expansions (with C. E. Jones and J. E. Young), *Phys. Rev. D* 4: 2358 (1971)

Current Conservation and Double-Spectral Representations for Scattering of Vector Particles (with M. Feigenbaum), *Phys. Rev. D* 4: 3738 (1972)

Generalized $O(2, 1)$ Expansions for Asymptotically Growing Amplitudes II: Space-Time Region (with C. E. Jones and J. E. Young), *Ann. Phys.* 70: 286 (1972)

Some General Consequences of Regge Theory for Pomeranchukon-Pole Couplings (with C. Edward Jones, S.-H. H. Tye, G. Veneziano, and J. E. Young), *Phys. Rev. D* 6: 1033 (1972)

Helicity Poles and the Triple Regge Vertex (with C. E. Jones and J. E. Young), *Phys. Rev. D* 6: 640 (1972)

Model of the Bare Pomeron, *Phys. Rev. D* 12: 163 (1975)

Classical Space-Time Concepts in High-Energy Collisions (with K. Gottfried), *Phys. Rev. D* 17: 2487 (1978)

Azimuthal Correlations in $e^+ e^-$ Jets: A Test of Quantum Chromodynamics (with So-Young Pi and R. L. Jaffe), *Phys. Rev. Lett.* 41: 142 (1978)

Variational Principles for Complex Eigenvalues (with C. E. DeTar), *Astrophys. J.* 227: 349 (1979)

Connection Between Quark-Model Eigenstates and Low-Energy Scattering (with R. L. Jaffe), *Phys. Rev. D* 19: 2105 (1979)

Behavior of Unstable Atoms in Applied Fields, *Phys. Rev. A* 20: 1567 (1979)

Tensor Analysis of Hadronic Jets in Quantum Chromodynamics (with J. F. Donoghue and So-Young Pi), *Phys. Rev. D* 20: 2759 (1979)

Quark Model States and Low Energy Scattering, Lectures delivered at the Erice School of Subnuclear Physics (July 1979). Published in *Pointlike Structures Inside and Outside Hadrons*, edited by A. Zichichi, Plenum, New York and London (1982)

Behavior of Decaying States in Applied Fields, *Physica* 96A: 260 (1979)

**Asymptotic Realms
of Physics**

Why the Renormalization Group Is a Good Thing

Steven Weinberg

My text for today is a paper by Francis Low and Murray Gell-Mann. It is "Quantum Electrodynamics at Small Distances," published in the *Physical Review* in 1954.

This paper is one of the most important ever published in quantum field theory. To give you objective evidence of how much this paper has been read, I may mention that I went to the library to look at it again the other day to check whether something was in it, and the pages fell out of the journal. Also it is one of the very few papers for which I know the literature citation by heart. (And all the others are by me.) This paper has a strange quality. It gives conclusions which are enormously powerful; it's really quite surprising when you read it that anyone could reach such conclusions: The input seems incommensurate with the output. The paper seems to violate what one might call the First Law of Progress in Theoretical Physics, the Conservation of Information. (Another way of expressing this law is: *You will get nowhere by churning equations.* I'll come to two other laws of theoretical physics later.)

I want here to remind you first what is in this paper, and try to explain why for so long its message was not absorbed by theoretical physicists. Then I will describe how the approach used in this paper, which became known as the method of the renormalization group, finally began to move into the center of the stage of particle physics. Eventually I will come back to the question in the title of my talk—why the renormalization group is a good thing. Why does it yield such powerful conclusions? And then at the very end, very briefly, I'll indicate some speculative possibilities for new applications of the ideas of Gell-Mann and Low.

Let's first take a look at what Gell-Mann and Low actually did. They started by considering an ancient problem, the Coulomb force between two charges, and they asked how this force behaves at very short distances. There's a naive argument that, when you go to a very high momentum transfer, much larger than the mass of the electron, the mass of the electron should become irrelevant, and therefore, since the potential has the dimensions (with $\hbar = c = 1$) of an inverse length, and since there is no other parameter in the problem with units of mass or length but the distance itself, the potential should just go like the reciprocal of the distance r. That is, you should have what is called naive scaling at very large momentum transfers or, in other words, at very short distances. Now, this doesn't happen, and this observation is the starting point of the paper by Gell-Mann and Low. The leading term in the potential, due to a one-photon exchange, is indeed just α/r. However, if you calculate the first radiative correction to the potential by inserting an electron loop in the exchanged photon line, you find a correction which has a logarithm in it:

$$V(r) = \frac{\alpha}{r}\left[1 + \frac{2\alpha}{3\pi}\left\{\ln\left(\frac{1}{\gamma m_e r}\right) - \frac{5}{6}\right\}\right] \quad (\gamma = 1.781\ldots). \tag{1}$$

This does not behave like $1/r$ as r goes to zero.

The questions addressed in the paper by Gell-Mann and Low are, first, why does the naive expectation of simple dimensional analysis break down? And, second, can we characterize the way this will happen in higher-order perturbation theory? And, third, what does the potential look like at really short distances, that is, when the logarithm is so large as to compensate for the smallness of $2\alpha/3\pi$? Those distances are incredibly short, of course, because α is small and the logarithm doesn't get big very fast. In this particular case the distance at which the logarithmic term becomes large is 10^{-291} cm. Nevertheless, the question of the behavior of the potential at short distances is an important matter of principle, one that had been earlier discussed by Landau and Källén and others, and that becomes also a matter of practical importance for forces that are stronger than electromagnetism.

Gell-Mann and Low immediately realized that the only reason that there can be any departure from a $1/r$ form for the potential is because the naive expectation that at large momentum transfer the electron mass should drop out of the problem is simply wrong. The potential does not have a smooth limit when r is very small compared to the Compton wavelength of the electron or, in other words, when the electron mass goes to zero. You can see from (1) that when the electron mass goes to zero the logarithm blows up. The failure of the naive expectation for the Coulomb force at short distances is entirely due to the fact that there is a singularity at zero electron mass. But where did that singularity come from? It's a little surprising that there should be a singularity here. In fact, if you look at the Feynman diagram in which an electron loop is inserted in the exchanged photon line, you can see that the momentum transfer provides an infrared cutoff and, in fact, there's no way that this diagram can have a singularity for zero electron mass. What is going on here?

Gell-Mann and Low recognized that the singularity at the electron mass is entirely due to the necessity of renormalization, in particular of what is called charge renormalization. If you calculate the one-loop diagram using an ultraviolet cutoff at momentum Λ to make the integral finite then the formula you get before you go to any limit is something like this (simplified a little bit):

$$V(r) = \frac{\alpha}{r}\left[1 + \frac{2\alpha}{3\pi}\ln\left(\sqrt{\frac{1 + r^2\Lambda^2}{1 + r^2 m_e^2}}\right) + \cdots\right]. \tag{2}$$

This is, as expected, not singular as the electron mass goes to zero. Consequently the naive expectation that the potential should go like $1/r$ at short distances is indeed correct for (2): the potential approaches α/r. The potential also behaves like $1/r$ at very large distances, but here with a different coefficient:

$$V(r) \to \frac{\alpha}{r}\left[1 + \frac{2\alpha}{3\pi}\ln\left(\frac{\Lambda}{m_e}\right) + \cdots\right] \quad \left(\text{for } r \gg \frac{1}{m_e} \gg \frac{1}{\Lambda}\right). \tag{3}$$

But the electric charge is *defined* in terms of this coefficient, because we measure charge by observing forces at large distances. That is, if we want to interpret α as the observed value of the fine-structure constant, then in (2) we should make the replacement

$$\alpha \to \alpha\left[1 - \frac{2\alpha}{3\pi}\ln\left(\frac{\Lambda}{m_e}\right) + \cdots\right], \tag{4}$$

so that (2) becomes (to second order in α)

$$V(r) = \frac{\alpha}{r}\left[1 - \frac{2\alpha}{3\pi}\ln\left(\frac{\Lambda}{m_e}\right) + \frac{2\alpha}{3\pi}\ln\left(\sqrt{\frac{1 + r^2\Lambda^2}{1 + r^2 m_e^2}}\right)\right]. \tag{5}$$

Now we can let the cutoff Λ go to infinity, and we get (1) (aside from nonlogarithmic terms, which are not correctly given by the simplified formula (2)). The singularity at zero electron mass arises solely from the renormalization (4) of the electric charge.

That is the diagnosis—now what is the cure? This too was provided by Gell-Mann and Low. They advised that since the logarithm of the electron mass was introduced by a renormalization prescription which defines the electric charge in terms of Coulomb's law at very large distances, we shouldn't do that; we should instead define an electric charge in terms of Coulomb's law at some arbitrary distance, let's say R; that is, we should define a renormalization-scale-dependent electric charge as simply the coefficient of $1/R$ in the Coulomb potential:

$$\alpha_R \equiv R V(R). \tag{6}$$

You might think that this wouldn't get you very far, but it does. Let's for a moment just use dimensional analysis, and not try to calculate any specific Feynman diagrams. If I set out to calculate the Coulomb potential at some arbitrary distance r, and I use as an input parameter the value of the fine structure constant α_R at some other distance R, then on dimensional grounds the answer must be a factor $1/r$ times a function of the dimensionless quantities α_R, r/R, and $m_e R$:

$$V(r) = \frac{1}{r} F\left(\alpha_R, \frac{r}{R}, m_e R\right). \tag{7}$$

Since we are expressing the answer in terms of α_R rather than $\alpha \equiv \alpha_\infty$, there should be no singularity at $m_e = 0$, and hence for r and R much less than $1/m_e$, the dependence on m_e should drop out here. Multiplying with r then gives our development equation for α:

$$\alpha_r = F\left(\alpha_R, \frac{r}{R}\right). \tag{8}$$

This is usually written as a differential equation $r d\alpha_r/dr = -\beta(\alpha_r)$, with $\beta(\alpha) \equiv -[\partial F(\alpha, x)/\partial x]_{x=1}$. However, it makes no difference in which form it is written; the important thing is that we have an equation for α_r in which $\alpha = 1/137$ and m_e do not enter, except through the initial condition that for $r = 1/m_e$, α_r is essentially equal to α.

This has remarkable consequences. First of all, one consequence which is not of stunning importance, but is useful: since $1/137$ and the mass of the electron only enter together, through the initial condition, you can relate the number of logarithms to the number of powers of $1/137$. For instance, we have seen that in the Coulomb potential to first order in $1/137$ there is only one logarithm, and it can be shown that in second order there's still only one logarithm, in third order there are two logarithms, and so on. That's interesting. It is surprising that one can obtain such detailed information about higher orders with so little work, but what is really remarkable is what (8) says about the very short distance limit. In the limit of very short distances, there are only two possibilities.

First α_R may not have a limit as R goes to 0, in which case the conclusion would be that the bare charge is infinite and probably (although I can't say this with any certainty) the theory makes no sense. Such a theory probably develops singularities at very short distances, like the so-called ghosts or tachyons, which make the theory violate the fundamental principles of relativistic quantum mechanics.

The second possibility is that α_R does have a limit as R goes to 0, and the limit is nonzero, but since 1/137 enters in this whole business just as the initial condition on (8), this limit is, of course, independent of 1/137. By letting r and R both go to zero in (8) with arbitrary ratio x, you can see that the limit α_0 of α_r as $r \to 0$ is defined as the solution of the equation

$$\alpha_0 = F(\alpha_0, x) \quad \text{(for all } x\text{)}. \tag{9}$$

This limit is called a fixed point of the development equation. (Another way of expressing this is that α_0 is a place where the Gell-Mann–Low function $\beta(\alpha)$ vanishes.)

The one thing which isn't possible in quantum electrodynamics is that the limit of α_r as $r \to 0$ should be 0. Although we can't calculate the development function in general, we can calculate it when α_r is small, so we can look and see whether or not, if α_r is small, it will continue to decreases as r goes to 0. The answer is no, it doesn't. When α_r is small, it's given by (1) as

$$\alpha_r = \alpha_R \left[1 + \frac{2\alpha_R}{3\pi} \ln \left(\frac{R}{r} \right) + \cdots \right]. \tag{10}$$

You see that when r gets very small α_r does not decrease, it increases. Eventually it increases to the point where you can't use the power series any more; this happens at a distance of 10^{-291} cm. About what happens at such short distances, this equation tells you essentially nothing, but the one thing it does tell you for sure is that when r goes to 0, α_r does not go to 0, because if it did go to 0 then you could use perturbation theory, and then you would see it doesn't go to 0; so it doesn't.

This analysis gives information about much more than the short-distance behavior of the Coulomb potential. Consider any other amplitude, let's say, for the scattering of light by light. This will have a certain dimensionality, let's say length to the dth power. So write this amplitude as the renormalization scale to the dth power times some dimensionless function of the momenta $k_1, k_2, \ldots$ of the various photons, the electron mass, the renormalization scale R, and the fine-structure constant α_R at that renormalization scale. (Remember the idea. We're defining the electric charge in terms of the Coulomb potential not at infinity but at some distance R.) That is, the amplitude A takes the form

$$A = R^d f(k_1 R, k_2 R, \ldots, m_e R, \alpha_R). \tag{11}$$

In order to study the limit in which $k_1 = k x_1$, $k_2 = k x_2$, $\ldots$ with x_1, x_2, $\ldots$ fixed and the overall scale k going to infinity, it is very convenient to choose $R = 1/k$. No one can stop you from doing that. You can renormalize anywhere you want; the physics has to be independent of where you renormalize. Now there is no singularity here at zero electron mass, because we renormalizing not at large distances but at short distances; hence we can replace $m_e R$ by 0 in the limit $R \to 0$. With $R = 1/k$ and $k \to \infty$, the amplitude (11) has the behavior

$$A \to k^{-d} f(x_1, x_2, \ldots, 0, \alpha_{1/k}). \tag{12}$$

The factor k^{-d} is what we would expect from naive dimensional analysis, ignoring problems of mass singularities or renormalization. Aside from this, the asymptotic behavior depends entirely on the behavior of the function α_r for $r \to 0$. In particular if α_r approaches a finite limit as $r \to 0$, then the amplitude does exhibit naive scaling for $k \to \infty$, but with a coefficient of k^{-d} that is not easy to calculate. (There are complications here that I have left out, having to do with matters like wavefunction renormalization. The above discussion is strictly valid only for suitably averaged cross sections. However the result of naive scaling for α_0 finite is valid for purely photonic amplitudes. For other amplitudes, there are corrections to the exponent d.)

Now this is really amazing—that one can get such conclusions without doing a lot of difficult mathematics, without really ever trying to look at the high orders of perturbation theory in detail. Nevertheless, the paper by Gell-Mann and Low suffered a long period of neglect—from 1954, when it was written, until about the early 1970s. There are a number of reasons for this; let me just run through what I think were the important ones.

First of all, there was a general lack of understanding of what it was that was important in the Gell-Mann–Low paper. There had been a paper written the year before Gell-Mann and Low, by Stueckelberg and Petermann, which made the same remark Gell-Mann and Low had made, that you could change the renormalization point freely in a quantum field theory, and the physics wouldn't be affected. Unfortunately, when the book on quantum field theory by Bogoliubov and Shirkov was published in the late 1950s, which I believe contained the first mention in a book of these matters, Bogoliubov and Shirkov seized on the point about the invariance with respect to where you renormalize the charge, and they introduced the term "renormalization group" to express this invariance. But what they were emphasizing, it seems to me, was the least important thing in the whole business.

It's a truism, after all, that physics doesn't depend on how you define the parameters. I think readers of Bogoliubov and Shirkov may have come into the grip of a misunderstanding that if you somehow identify a group that then you're going to learn something physical from it. Of course, this is not always so. For instance when you do bookkeeping you can count the credits in black and the debits in red, or you can perform a group transformation and interchange black and red, and the rules of bookkeeping will have an invariance under that interchange. But this does not help you to make any money.

The important thing about the Gell-Mann–Low paper was the fact that they realized that quantum field theory has a scale invariance, that the scale invariance is broken by particle masses but these are negligible at very high energy or very short distances if you renormalize in an appropriate way, and that then the only thing that's breaking scale invariance is the renormalization procedure, and that one can take that into account by keeping track of the running coupling constant α_R. This didn't appear in the paper by Stueckelberg and Petermann, and it was pretty well submerged in the book by Bogoliubov and Shirkov. I say this with some bitterness because I remember around 1960 when that book came out thinking that the renormalization group was pretty hot stuff, and trying to understand it and finding it just incomprehensible and putting it away. I made the mistake of not going back and reading carefully the paper by Gell-Mann and Low, which is quite clear and explains it all very well. (Incidentally, the later textbook by Bjorken and Drell gave a good clear explanation of all this, following the spirit of the Gell-Mann–Low paper.)

The second reason, I think, for these decades of neglect of the Gell-Mann–Low paper was the general distrust of quantum field theory that set in soon after the brilliant successes of quantum electrodynamics in the late 1940s. It was realized that the strong interactions were too strong to allow the use of perturbation theory and the weak interactions did not seem to have the property that the electromagnetic interactions did, of being renormalizable. (Renormalizability means that you can have a Lagrangian or a set of field equations with a finite number of constants, and all the infinities can always be absorbed into a redefinition of the constants, as I've already shown here that you can do with the cutoff dependence of the Coulomb potential.) Since people were not all that enthusiastic about quantum field theory, it was not a matter of high priority to study its properties at very short distances. Finally, we have seen, in quantum electrodynamics the Gell-Mann–Low analysis itself tells you that perturbation theory fails at very short distances, and then you just have to give up. There didn't seem to be much more that one could do.

The great revival of interest in the renormalization group came in the early 1970s, in part from a study of what are called anomalies. Anomalies are things that happen in higher orders of quantum field theory that you don't expect and that don't appear when you use the field equations in a formal way. I guess you could say the anomalies represent an instance of the Second Law of Progress in Theoretical Physics, which can be stated: *Do not trust arguments based on the lowest order of perturbation theory.* Some of these anomalies were studied here at MIT by Jackiw and Bell and Johnson and Low, and at Princeton by Steve Adler. In 1971 Callan, Coleman, and Jackiw were studying the scaling behavior of higher-order contributions to scattering amplitudes, and found as Gell-Mann and Low had found earlier in a different context that these amplitudes did not have the sort of "soft" nonsingular dependence on particle masses as the lowest-order contribution. A little later, Coleman and Jackiw traced this failure of naive scaling to an anomaly in the trace of the energy-momentum tensor. In the limit of zero masses one would expect the trace of the energy-momentum tensor to vanish. (For hydrodynamics, for instance, the trace of the energy-momentum tensor is 3 times the pressure minus the density. And everyone knows that for massless particles like light, the pressure is 1/3 the energy density. So you should get zero.) And, in fact, in quantum electrodynamics you do get zero if you just use lowest-order perturbation theory, in the limit where the electron mass is zero—but even with the electron mass equal to zero, if you calculate matrix elements of the energy-momentum tensor beyond the lowest-order perturbation theory you find that its trace is not zero. At about the same time, Callan and Symanzik set up a formalism for studying the failure of naive scaling. Their results turned out to look very much like the Gell-Mann–Low formalism. With the benefit of hindsight, this should not be surprising at all because, as I have emphasized here, the essential point of Gell-Mann and Low was that naive dimensional analysis breaks down precisely because of renormalization. The fact that the Coulomb potential is not just proportional to $1/r$ at short distances is one symptom of this breakdown of scale invariance, and the nonvanishing of the trace of the energy-momentum tensor is another symptom. The formalism used for one is related to the formalism used for the other.

Another theoretical influence: in the early 1970s non-Abelian gauge theories began to be widely studied, both with regard to the electroweak interactions and soon also with regard to the strong interactions. Politzer and Gross and Wilczek realized that the plus sign in the logarithmic term of (10), which prevented the use of perturbation theory in quantum electrodynamics at short distances, for non-Abelian gauge theories is a minus sign. The important thing about non-Abelian gauge theories for these purposes is that instead of one photon you have a family of "photons," and each member of this family of "photons" carries the "charge" that other members interact with. The prototypical non-Abelian gauge theory is that of Yang and Mills, in which there are three "photons." Because the "photons" interact with "photons," in addition to the usual diagrams for the "Coulomb" potential where you have loops of fermions like electrons inserted into exchanged "photon" lines, here you also have "photon" loops, and these have opposite sign. In fact, not only do they have opposite sign but they're bigger. In place of the characteristic factor of $2/3$ in (10), each "photon" loop carries a factor of $-11/3$. In the theory of strong interactions the fermions are quarks and there are 8 "photons" known as gluons. So unless you have an awful lot of quarks, the gluons are likely to overpower the quarks and give the logarithm in (10) a large negative coefficient, while in quantum electrodynamics you find a positive one. This makes all the difference because it means that as you go to shorter distances the forces get weak rather than getting strong and you can then use perturbation theory at very short distances. This is called asymptotic freedom. Politzer, Gross, and Wilczek instantly realized that this explains an experimental fact which had been observed in a famous experiment on deep inelastic electron proton scattering done by an MIT-SLAC collaboration in 1968. This was that at very high momentum transfer, in other words, at very short distances, the strong interactions seem to turn off and the formulas for the form factors in deep inelastic electron scattering seem to obey a kind of naive scaling, "Bjorken scaling." This had been a mystery because it would require that somehow or other the strong interactions must disappear at short distances. It had been this result that in part had stimulated all this theoretical work on scaling. Now suddenly this was understood.

Also, if the force gets small as one goes to short distances, there's a least a good chance that it will get big as you go to large distances. At first it was generally supposed that this did not happen. It was assumed that the "photons" here are heavy, getting their mass (like the intermediate vector bosons of the weak interactions) from the vacuum expectation values of scalar fields. But scalar fields would have raised all sorts of problems for the theory. Then Gross and Wilczek and I guessed that there are no strongly interacting scalars; that the gluons, the strongly interacting "photons," are therefore massless; that consequently the force does continue to increase with distance; and that this might explain why we don't observe the gluons, and also why we don't observe the quarks. Putting together all the pieces, at last we had a plausible theory of the strong interactions. It was christened (by Murray Gell-Mann, who with Fritsch and Minkowski had developed some of these ideas before the discovery of asymptotic freedom) quantum chromodynamics, that is, the same as quantum electrodynamics except that the quantity called color replaces electric charge.

There's an interesting side to the history of all this. Ken Wilson, perhaps alone of all theoretical physicists, was well aware of the importance of using the renormalization group ideas of Gell-Mann and Low through the late 1960s and early 1970s. He used these ideas to consider all kinds of interesting things that might happen at high energy. He considered, for example, the possibility that the coupling constant would go to a nonzero fixed point, which is exactly what Gell-Mann and Low thought might happen in quantum electrodynamics, or that we might find a limit cycle where the coupling constant goes round and round and just keeps oscillating in a periodic way. He wrote papers about how this would appear from various points of view experimentally, whereas the experimentalists at the same time were showing that, in fact, everything is very simple—that at high energies the strong interactions go away altogether. To the best of my knowledge, Ken Wilson missed only one thing—the possibility that the coupling constant might go to zero at short distances. He just didn't consider that possibility because he knew it didn't happen in quantum electrodynamics. On the other hand, Tony Zee was very much aware of that possibility, and wrote a paper saying, wouldn't it be simply grand if the coupling constant did go to zero at high energy, then we could understand the MIT-SLAC experiment. He sat down and calculated the logarithmic terms in the vacuum polarization effect in various theories and he found he got the plus sign, the one that you get in quantum electrodynamics, in all his calculations, and gave up in disgust. The one case he did not consider was the case of a non-Abelian gauge theory like the Yang-Mills theory. The reason that he didn't consider it was because at that time the rules for calculating those theories, with Fadeev-Popov ghosts and all the rest of the boojums, were not very widely known and he didn't feel confident in doing the calculation. So he gave up the idea. On the other hand, Gerard 't Hooft, who knows every-thing about how to calculate in a non-Abelian gauge theory, did this calculation and, in fact, found that the sign factor in the Gell-Mann–Low function was opposite to what it is in quantum electrodynamics. He announced the result of this calculation at a conference on gauge theory at Marseille in June 1972, but he waited to publish it while he was doing other things, so his result did not attract much attention.

Finally, however, it did all come together. From 1973 on, I would say, most theorists have felt that we now understand the theory of the strong interactions. It is, of course, very important to test this understanding, and I certainly wouldn't claim that quantum chromodynamics is indisput-ably verified. My own feeling is that quantum chromodynamics *will* be indisputably verified in machines like LEP, in which electron-positron annihilation produces jets of quarks and antiquarks and gluons, and that this verification will be very much like the verification of quantum electrodynamics, not in the 1940s when the problem was the loop graphs, but in the 1930s when quantum electrodynamics was verified for processes like Bhabha scattering and Møller scattering and Compton scattering, using only tree diagrams. I say this in part because of a theorem, that if you calculate the cross section not for producing a certain definite number of quarks or gluons but instead for producing a certain definite number of quark or gluon jets (a jet being defined as a cone within which there can be any number of particles) then these cross sections satisfy the assump-tions of the Gell-Mann–Low paper, that in the limit of very short distance or very high energy they can simply be calculated by perturbation theory. The other case in which one would like to verify quantum chromodynam-ics is, of course, at large distances or low energy, where the question of quark trapping arises. We'd like to be able to calculate the mass of the proton, the pion-nucleon scattering at 310 MeV, and all sorts of other quantities. Many people are working on this very difficult problem. I will come back at the end of my talk to one idea about how this kind of calculation might be done.

The wonderful discovery by Politzer, Gross, and Wilczek of the decrease of the strong interactions at high energy also had an immediate impact on our understanding of the possibilities for further unification. Ideas about unifying the strong and electroweak interactions with each other have been presented in papers by Pati and Salam, Georgi and Glashow, and many others. However, there was from the start an obvious problem with any such idea: strong interactions are strong and the others aren't. How can you unify interactions that have such different coupling constants? Once quantum chromodynamics was discovered, the possibility opened up that because the strong interactions, although strong at ordinary energies, get weak as you go to high energy or short distances, at some very high energy they fuse together with the electroweak interactions into one family of "grand unified" interactions. This idea was proposed in 1974 by Georgi, Quinn, and me, and we used it to calculate the energy at which the strong and electroweak couplings come together. After my earlier remarks, it should come as no surprise to you that the energy that we found is an exponential of an inverse square coupling constant, like the energy that Gell-Mann and Low found where electromagnetism would become a strong interaction. (They expressed this in terms of distances, but it's the same thing, except for taking a reciprocal.) Instead of the Gell-Mann–Low energy of $\exp(3\pi/2\alpha)$ electron masses, we found that (in a large class of theories) the strong and electroweak forces come together at an energy which is larger than the characteristic energy of quantum chromodynamics by a factor roughly $\exp(\pi/11\alpha)$. (The 11 is that magic number I mentioned earlier that is always contributed by a loop of gauge bosons.) This factor, in other words, is something like the 2/33 power of the enormous factor that corresponds to the incredibly short distance at which Gell-Mann and Low found that perturbation theory begins to break down in quantum electrodynamics. The energy here turns out to be something still very high but not so inconceivably high, only about 10^{15} GeV. This suggests that there's a whole new world of physics at very high energies of which we in studying physics at 100 GeV or thereabouts are only seeing the debris.

There may be all sorts of new physical effects that come into play at 10^{15} GeV. For example, there's no real reason to believe that baryon number would be conserved at such energies. The fact that it is conserved at ordinary energy can be understood without making any assumption about baryon conservation as an exact symmetry of nature. We might expect a proton lifetime of the order of magnitude of $(10^{15}\text{ GeV})^4/(\alpha^2 m_p^5)$, essentially as estimated in the paper by Georgi, Quinn, and me. This comes out to be about 10^{32} years, which is nice because it's a little bit beyond the lifetimes that have been looked for so far experimentally, but not hopelessly beyond them. Of course, we are all anxious to find out whether or not the proton does decay with some such lifetime.

After the strong and electroweak interactions have hooked up with each other, what happens then? Does the grand unified interaction, which then would have only one independent coupling constant, satisfy the idea of asymptotic freedom, that the coupling constant goes on decreasing? Or does the coupling start to rise, presenting us back again with the same problem that Gell-Mann and Low faced, of a coupling constant which increases as you go to short distances or high energies and, therefore, ultimately makes it impossible to use perturbation theory. And, of course, at 10^{15} GeV you're not very far below the energy at which gravity becomes important. Perhaps that cancels all bets.

In addition to the applications of the renormalization group to the strong interactions and thence to grand unified theories, there had even a little earlier been an entirely different development due to Ken Wilson and Michael Fisher and Leo Kadanoff and others—the application of renormalization-group methods to critical phenomena. It is interesting that in this volume there are two papers that deal with fixed points and the renormalization group and so on. The first of these, by Ken Johnson, is entirely about quantum field theory. The second one, by Mitchell Feigenbaum, is entirely about statistical phenomena. In fact, there seems to be no overlap between these papers except for the language of the renormalization group.

I think it is really surprising that the same ideas can be applied to such apparently diverse realms. When you're dealing with critical phenomena, you're not concerned about short-distance (or high-energy) behavior; you're concerned about long-distance behavior. You're asking about matters like critical opalescence, about the behavior of the correlation function when two points go to very large separation, not very short separation. Well, that alone is perhaps not such an enormous difference. After all even in quantum electrodynamics you might be interested in such questions, not in the real world where the electron mass provides an infrared cutoff which makes all such questions irrelevant, but say in a fictitious world where the electron mass is zero. If the electron mass really were zero, it would be very interesting to say what happens to quantum electrodynamics at very long distances. The Gell-Mann–Low formalism answers that question. At very long distance, massless quantum electrodynamics becomes a free field theory. In quantum chromodynamics all we know for sure is that it does *not* become a free field theory, just the reverse of what we know about the short-distance behavior of these theories. Now, when you're talking about critical phenomena there is something analogous to the mass of the electron—there's the difference between the actual temperature and the critical temperature. The critical temperature at which a second-order phase transition occurs is defined in such a way that at that temperature there's nothing that's providing an infrared cutoff, and, therefore, for example, correlation functions don't exponentially damp as you go to very large separations. So, in other words, setting the temperature equal to the critical temperature in a statistical mechanics problem is analogous to studying what happens in quantum electrodynamics when you actually set the electron mass equal to zero and then consider what happens as you go to very large distances. Of course, we can't dial the value of the electron mass. We can, however, set thermostats, so there are things that are of interest in statistical mechanics that aren't of that much interest in quantum field theory, because the value of the temperature really is at our disposal. When you look at it from this point of view you can see the similarity between what people who work in critical phenomena are doing and what Gell-Mann and Low did. They're all exploiting the scale invariance of the theory, scale invariance, that is, except for the effects of renormalization, and corrected by the Gell-Mann–Low formalism for the effects of renormalization.

There is another difference between high energy particle physics and statistical physics. After all, ordinary matter is, in fact, not scale invariant. Where does scale invariance come from when you're talking about critical opalescence in a fluid going through a phase transition? In what sense is there any scale invariance with or without renormalization corrections? If you construct a kind of field theory to describe what's happening in a fluid, in which the field ϕ might be a pressure or density fluctuation of some kind, the Hamiltonian would include a huge number of terms, ϕ^2, ϕ^4, ϕ^6, ... because there's no simple principle of renormalizability here that limits the complexity of the theory. It doesn't look like a scale-invariant theory at all. Well, in fact, you can show that if you're interested in the long-distance limit then all the higher terms such as ϕ^6, ϕ^8, etc., become irrelevant. The ϕ^2 term also would break scale invariance, but this is precisely the effect we eliminated by going to the critical temperature. Finally, the ϕ^4 term also breaks scale invariance (in classical statistical mechanics its coupling constant has the dimensions of a mass); but this is taken care of by the same renormalization group manipulations that are needed anyway to deal with renormalization effects. If C is a function with dimensionality d that describes correlations at separation r, then at the critical temperature dimensional analysis gives

$$C = R^d F(r/R, R\lambda(R)), \tag{13}$$

where $\lambda(R)$ is the ϕ^4 coupling constant, defined by some renormalization prescription at a scale R. Once again, set R equal to r; (13) then becomes

$$C = r^d F(1, r\lambda(r)). \tag{14}$$

Furthermore, the dimensionless quantity $r\lambda(r)$ satisfies a Gell-Mann–Low equation like (8). If this quantity approaches a fixed point for $r \to 0$, then (14) indicates that we have naive scaling ($C \propto r^d$) for $r \to 0$. (Once again, I am ignoring complications having to do with the renormalization of the field ϕ, or equivalently of the operator $\partial_\mu \phi \partial^\mu \phi$. These change the value of the power of r as $r \to 0$.)

There is still something mysterious about all this, which takes me back to my starting question: Why is the renormalization group a good thing? What in the world does renormalization have to do with critical phenomena? Renormalization was invented in the 1940s to deal with the ultraviolet divergences in quantum field theory. Theories of condensed matter are not renormalizable field theories. They don't look like quantum electrodynamics at all. If you throw away the higher terms in the Hamiltonian (ϕ^6, ϕ^8, etc.) on the grounds that you're only interested in long-distance behavior (these terms are what in statistical mechanics are called irrelevant operators), then you're left with a theory that doesn't have any need for renormalization to eliminate ultraviolet divergences. (This is because when you deal with critical phenomena you're working with 3 and not 4 dimensions.) But then why does the use of the renormalization group help at all in understanding critical phenomena?

I think the answer to the last question gets us to essence of what really is going on in the use of the renormalization-group method. The method in its most general form can I think be understood as a way to arrange in various theories that the degrees of freedom that you're talking about are the relevant degrees of freedom for the problem at hand. If you renormalize in the conventional way in quantum electrodynamics in terms of the behavior of the Coulomb potential at large distances, then for any process like scattering of light by light you will have momenta running around in the Feynman diagram which go down to small values, small meaning of the order of the electron mass. Even if what you're really interested in is the scattering of light by light at 100 GeV, when you calculate the Feynman diagram you'll find that the integrals get important contributions from momenta which go all the way down to one-half MeV, the electron mass. In other words, the conventional renormalization scheme in quantum electrodynamics, although it does not actually introduce any mistakes, emphasizes degrees of freedom which, when you're working at very high energy, are simply not the relevant degrees of freedom. The Gell-Mann–Low trick of introducing a sliding renormalization scale effectively suppresses those low-energy degrees of freedom in the Feynman integrals. If you define a renormalization scheme, so that when you calculate scattering of light by light at 100 GeV you use a definition of the electric charge which is renormalized at 100 GeV, then you will in fact find that all of the Feynman integrals you have to do get their important contributions from energies roughly of order 100 GeV. In other words, the Gell-Mann–Low procedure gets the degrees of freedom straight. The same is true in the renormalization-group approach to critical phenomena, whether you implement it as Wilson did by simply integrating out the very short wave numbers, or if you do what Brezin, LeGuillou, and Zinn-Justin do and use the renormalization scheme itself to provide an ultraviolet cutoff in close analogy to the Gell-Mann–Low approach to field theory. Either way, you are arranging the theory in such a way that only the right degrees of freedom, the ones that are really relevant to you, are appearing in your equations. I think that this in the end is what the renormalization group is all about. It's a way of satisfying the Third Law of Progress in Theoretical Physics, which is that *you may use any degrees of freedom you like to describe a physical system, but if you use the wrong ones, you'll be sorry.*

Now let me briefly come to some possibilities for future developments. We still have with us the problem of quantum chromodynamics at very large distances. This is a somewhat paradoxical problem because in fact for a long time we have had a perfectly good quantum field theory for strong interactions at very large distances. For simplicity, I will adopt here the fiction that the bare quark masses are zero, which for many purposes is a good approximation. In that case, the pion is massless because it's a Goldstone boson. The Lagrangian that describes strong interactions at a very low energy like 1 eV, where the only degree of freedom is the massless pion, is the nonlinear Lagrangian which was originally written down by Gell-Mann and Levy in 1960, and which as I showed in 1967 actually reproduces all the theorems of current algebra. The Lagrangian is

$$\mathscr{L} = -\partial_\mu \boldsymbol{\pi} \cdot \partial^\mu \boldsymbol{\pi} / (1 + \pi^2/F_\pi^2)^2, \tag{15}$$

where $\boldsymbol{\pi}$ is the pion field, and F_π is an empirically determined constant, about 190 MeV. This then is the field theory of the strong interactions at very low energy, always with the proviso that the bare quark masses are zero. (It's not much more complicated otherwise.) So we have a perfectly good field theory for strong interactions at low energies, and we also have a perfectly good field theory for strong interactions at very high energies, the quantum chromodynamics in which we all believe. The question is not so much how we can solve the strong interactions at low energy, or at large distances, as how we can prove that there's any connection between these two theories. How can we prove that if you start with quantum chromodynamics which we think is, in some sense, an underlying theory, that then if you then treat it in the limit of very long distances or low energies you go over to the theory described by (15)?

I wonder if the answer is not that we should expand once again our idea of what the renormalization group means. To me the essence of the renormalization-group idea is that you concentrate on the degrees of freedom that are relevant to the problem at hand. As you go to longer and longer wave lengths you integrate out the high-momentum degrees of freedom because they're not of interest to you and then you learn about correlation functions at long distances; or, vice versa, you do what Gell-Mann and Low did, and as you go to shorter and shorter wave lengths you suppress the long wavelengths. But sometimes the choice of appropriate degrees of freedom is not just a question of large or small wavelength, but a question of what kind of excitation we ought to consider. At high energy the relevant particles are quarks and gluons. At low energy they're massless pions. What we need is a version of the renormalization group in which as you go from very high energy down to low energy you gradually turn on the pion as a collective degree of freedom, and turn off the high-energy quarks. Now I don't really know how to do that. I do have some ideas about it. There are ways of introducing fields for particles like the pion which are not elementary, and then making believe that they are elementary. The question is whether the dynamics generate a kinematic term for π in the Lagrangian. I'm working on this and certainly have no progress to report. I have asked my friends in statistical mechanics whether or not when they use renormalization-group ideas they find that they have to not only continually change the wavelength cutoff but actually introduce new degrees of freedom as they go along. Apparently this has not been done in statistical mechanics. Collective degrees of freedom, like say the Cooper pair field in superconductivity, are just introduced at the beginning of the calculation and are not turned on in a smooth way as you go to long wavelengths. But perhaps this readjustment of degrees of freedom might be useful also in statistical mechanics.

Finally, I want to come to what is perhaps the most fundamental question of all: What is the behavior of nonrenormalizable theories at short distances? This is an important problem above all because so far no one has succeeded in embedding gravity into the formalism of a renormalizable quantum field theory. As far as we know, the Lagrangian for gravity, in order to cancel all infinities, has to be taken to have an infinite number of terms, in fact all conceivable terms which are allowed by general covariance and other symmetries. For instance, for pure gravity the Lagrangian must be taken as

$$\mathscr{L} = \frac{1}{16\pi G} R + fR^2 + f' R^{\mu\nu} R_{\mu\nu} + hR^3 + \cdots . \qquad (16)$$

(I've written here only terms involving the metric but in reality there are an infinite number of terms involving matter as well.) This is not at all in contradiction with experiment; the success of Einstein's theory does not contradict this. The leading term, the R term, has a coefficient of about 10^{38} GeV^2; that is the square of the Planck mass. If we believe that this is the only unit of mass in the problem then the coefficients f and f' in the next two terms are of order 1; the coefficients h, etc., in the next few terms are of order 10^{-38} GeV^{-2}; and so on. Any experiment which is carried out at distances large compared to 10^{-19} GeV^{-1} (which, of course, all experiments are) would only see the R term. So we don't know anything experimentally about the higher terms in (16). There's no evidence for or against them except that if gravity isn't renormalizable they would all have to be there.

What would be the short-distance or the high-energy behavior of such a theory? Well, suppose we make a graph in coupling-constant space showing the trajectory of the coupling constants G, f, f', h, etc., as we vary the renormalization scale. The renormalization group applies here; a theory doesn't have to be renormalizable for us to apply the renormalization-group method to it. These trajectories simply describe how all the couplings change as you go from one renormalization scale to another. Now many of those trajectories—in fact, perhaps most of them—go off to infinity as you go to short-distance renormalization scales. However, it may be that there's a fixed point somewhere in coupling-constant space. A fixed point, remember, is defined by the condition that if you put the coupling constant at that point it stays there as you vary the renormalization scale. Now, it is a fairly general phenomenon that for each fixed point there are usually some trajectories that hit the point, but these trajectories do not fill up much of coupling-constant space. That is, there may be some trajectories that you can draw that run into a given fixed point, but the surface that these trajectories map out is usually finite-dimensional, whether the theory has an infinite number of couplings or not.

There's even experimental evidence for this property of fixed points. In fact, the whole lore of second-order phase transitions in a sense can be quoted as experimental evidence for this statement. In second-order phase transitions, where you're considering not the behavior at short distances but at large distances, this statement translates into the statement that the *normals* to the surfaces of trajectories (now going the other way!) that hit a given fixed point form a finite-dimensional set. That is why in statistical mechanics, for example, if you want to produce a second-order phase transition, you only have to adjust one or a few parameters, so that the coupling constants have no components along these normals. Water is an extremely complicated substance, with a huge number of parameters describing all its molecules, but if you want to produce a second-order phase transition in water, all you have to do is adjust the temperature and the pressure; you don't also have to adjust the mass of the water molecules or the various force constants. This means that the surface formed by the trajectories which are attracted by the fixed point as you go to very long distances has only two independent normals. If you go to short distances instead, then that statement translates into the statement that the space of trajectories that are attracted to the fixed point is only 2-dimensional.

If the parameters of a theory lie on a trajectory that hits a fixed point at short-distance renormalization scales, then the physical amplitudes of the theory may be expected to behave smoothly at short distances or high energies—often just a power-law behavior, perhaps with anomalous exponents. The behavior of such a theory is just like that found by Gell-Mann and Low for quantum electrodynamics with an ultraviolet fixed point. On the other hand, one may suspect that a theory which is on a trajectory which does *not* hit any fixed point is doomed to encounter a Landau ghost or a tachyon or some other terrible thing. Then you have a reason for believing that nature has to arrange the infinite number of parameters in a nonrenormalizable field theory like the theory of gravity so that the trajectories do hit the fixed point. This would leave only a finite number of free parameters. Indeed, conceivably this finite-dimensional surface is only 1-dimensional—conceivably it's just a line running into the fixed point. In this case we would have a physical theory in which the demands of consistency, the demands of unitarity and analyticity and so on which rule out ghosts and tachyons, dictate all the parameters of the theory, except for one scale parameter which just specifies the unit of length. What could be better?

The Physics of Asymptotic Freedom

Kenneth A. Johnson

I. Introduction

I would like to dedicate this talk to Francis Low on the occasion of his sixtieth birthday. I have been fortunate to have had the pleasure of having had the office next to his for longer than probably either of us (certainly I) care particularly to think about. I have been fortunate because Francis is one of those rare persons who listens with close attention to what is said to him. He immediately recognizes all the shortcomings and logical problems. He is also very helpful in suggesting how these might be overcome or not, as the case might be. We miss him a lot in the Center for Theoretical Physics.

I hope that Francis will find amusing this short review of the work of others which concerns the physics of "asymptotic freedom." To make it complete for other readers, I have also taken the liberty of giving a brief resumé of some of the relevant aspects of the seminal work of Murray Gell-Mann and Francis Low from 1954 on "Quantum Electrodynamics at Short Distances." [1]

In that study Gell-Mann and Low provided an almost definitive analysis of the local structure of the Green's functions of renormalizable field theories. In accomplishing this, the scale-dependent (or "running") coupling constant plays a key role. It is this feature which I would like to discuss here in the context of non-Abelian or Yang-Mills field theories. In particular, I would like to advertise some recent work on the physical meaning of the so-called asymptotic freedom of these field theories. Asymptotic freedom, or the decrease in the effective coupling parameter with increasing momentum transfer, is the feature which has played the major role in motivating the study of gauge theories as serious candidates for describing the strong interaction between the fundamental particles. In particular, the experimental discovery of scaling in ep scattering [2], and the subsequent very successful phenomenological description of the scattering amplitude using the parton model [3], motivated a field theoretic "derivation" of the parton model. The great theoretical discovery [4] was that this derivation can only be achieved if the partons (i.e., the quarks) interact by means of a non-Abelian gauge field.

Because of the uniqueness of non-Abelian gauge field theories in this respect, many have made an effort to give a simple physical explanation of asymptotic freedom [5]. All of these have focused on the "electric" aspects of the gauge field. Here I shall look at the "magnetic" field to explain it. Everything I will discuss is generally true in arbitrary gauge theories. For simplicity, and so that I can use familiar terminology, I shall focus on electrodynamics. Thus, the non-Abelian SU (2) gauge theory will be considered as an "Abelian" gauge theory, that is, the quantum electrodynamics of massless charged spin-1 particles. Although the renormalizability of this theory to all orders hinges crucially on the fact that the theory is also symmetric under gauge transformations which mix the photon with the charged particles, for asymptotic freedom to lowest order, this plays no role. Asymptotic freedom will be shown to be a simple physical consequence of the spin of the field.

The physical insight on which I am reporting here was developed in an unpublished preprint by N. K. Nielsen [7]. I learned about this work from physicists at the Niels Bohr Institute a little over a year ago. Closely related papers have been written independently by R. J. Hughes [8].

In the final section, I shall briefly discuss some research [9] in progress that has been motivated partly by the intuition gained from the "magnetic" explanation of asymptotic freedom.

II. The Running Coupling in Quantum Electrodynamics [1]

The interaction between charges e_0 in the absence of matter is described by the potential (momentum space)

$$\frac{e_0^2}{q^2},\tag{2.1}$$

where q is the momentum transfer. In the presence of matter (in this case the vacuum matter consisting of charged fields) it becomes

$$\frac{e_0^2}{q^2 \varepsilon(q)},\tag{2.2}$$

where $\varepsilon(q)$ is the "dielectric" function. In field theory, e_0 and ε are cutoff (Ω) dependent quantities. For $q \gg \Omega$ we have $\varepsilon \to 1$, and the "bare" charge e_0 describes the interaction. One may define a "running" charge $e(q)$ so that

$$e^2(q) = \frac{e_0^2}{\varepsilon(q)}. \tag{2.3}$$

Here $e(q)$ is equivalent to an effective charge in the absence of matter which would describe the interaction at momentum transfer q.

Since the Hamiltonian of the field theory is simply expressed in terms of e_0, one might imagine computing $\varepsilon(q)$ perturbatively as a function of e_0. One then finds that everything so expressed has an explicit cutoff dependence. In "renormalizable" field theories, if one instead expresses $\varepsilon(q)$ in terms of $e(\lambda)$, where $q^2 = \lambda^2$ is a spacelike momentum transfer, then for q^2, $\lambda^2 \ll \Omega^2$, all quantities are insensitive to the cutoff Ω and the way in which it is effected. To all orders of $e^2(\lambda)$, $e^2(q)$ is given in terms of $e^2(\lambda)$ in a finite way when $\Omega \to \infty$.

It is usual in electrodynamics to take the value of λ to be zero, that is, to express the theory in terms of the long distance effective charge $e(0)$. The crucial observation of Gell-Mann and Low was that if one is interested in the form the theory takes at short distances (short in comparison to the scale set by the masses of the charged particles) this may be achieved in a simple form only if the charge $e(0)$ is replaced by $e(\lambda)$, where λ is also large in comparison to all particle masses. Thus they argued that renormalizability, as well as implying an insensitivity to the ultraviolet cutoff Ω, also implies that $e^2(q)$ given in terms of $e^2(\lambda)$ is insensitive to all charged-particle masses m, when q^2, $\lambda^2 \gg m^2$. It is this crucial observation that has proved to be useful. The reason for this is that the Feynman graphs which determine $e^2(q)$ in terms of $e^2(0)$ in general diverge in the infrared when the charged particle masses are set to zero. However, $e^2(q)$, when expressed in terms of $e^2(\lambda)$, has no such sensitivity to masses since λ serves to cut off all infrared divergent integrals. That is, the sensitivity to masses in renormalizable field theories is always a consequence of using the long-distance effective charge to parameterize the interaction.

More formally, Gell-Mann and Low showed that $e^2(q)$ is given in terms of $e^2(\lambda)$, when q and λ are large in comparison to all masses, by the equation

$$e^2(q) = e^2(\lambda) + \psi(e^2(\lambda)) \ln\left(\frac{q^2}{\lambda^2}\right) + O\left(\left(\ln\frac{q^2}{\lambda^2}\right)^2\right), \tag{2.4}$$

where $\psi(x) = x^2\psi_0 + x^3\psi_1 + \cdots$, and where ψ_0, ψ_1, etc., are numerical constants independent of any parameters. Here $\psi(x)$ is the famous Gell-Mann and Low function. The dependence of $e^2(q)$ on $e^2(\lambda)$ comes only through a constant of integration in the differential equation

$$q^2 \frac{\partial}{\partial q^2} e^2(q) = \psi(e^2(q)), \tag{2.5}$$

which is equivalent to (2.4), and can be obtained from it by differentiating $e^2(q)$ with respect to q^2, and then setting $\lambda^2 = q^2$.

It should be noted that the introduction of λ and $e^2(\lambda)$ may also be regarded as a particular way of determining $e^2(q)$ in terms of a "cutoff" λ, where $e^2(\lambda)$ is the corresponding "bare" coupling constant. Thus, in the domain $\lambda^2 > q^2 \gg m^2$, where λ is regarded as a cutoff,

$$e^2(q) = \frac{e^2(\lambda)}{\varepsilon_\lambda(q)}, \tag{2.6}$$

where $\varepsilon_\lambda = 1$ at $q^2 = \lambda^2$. Viewed in this way, the "dielectric" constant is

$$\frac{1}{\varepsilon_\lambda(q)} = 1 - \frac{\psi(e^2(\lambda))}{e^2(\lambda)} \ln\left(\frac{\lambda^2}{q^2}\right) + \cdots. \tag{2.7}$$

Since vacuum "matter" looks the same in all Lorentz systems, the "magnetic" permeability μ is given by $\varepsilon\mu = 1$, so

$$\mu_\lambda(q) = 1 - \frac{\psi(e^2(\lambda))}{e^2(\lambda)} \ln\left(\frac{\lambda^2}{q^2}\right) + \cdots. \tag{2.8}$$

Thus the magnetic susceptibility χ is negative (diamagnetism) when $\psi > 0$, which is the ordinary circumstance of electric screening ($\varepsilon > 1$).

Further, when $\psi < 0$, and there is asymptotic freedom (decreasing $e^2(q)$ as q increases), there is paramagnetism and antiscreening. This is the usual circumstance of the Yang-Mills theory. Here I should like to give a simple derivation of the sign (and magnitude) of ψ to lowest order which will reveal the simple physical origin of the destinction between particles which screen and particles which antiscreen.

Although I have reviewed the method of Gell-Mann and Low of introducing a running coupling $e^2(q)$ which depends upon an arbitrary scale λ at which the coupling is $e^2(\lambda)$, a scale could also have been introduced in another way through the use of a background field. Consider the vacuum in the presence of a uniform background magnetic field. The use of a background magnetic field rather than a background electric field is important for two reasons. First, the use of a magnetic field B, limits the size of vacuum currents to orbits with area of order $1/eB$ even when the particles are massless. This is sufficient to eliminate infrared divergences (however, for spin $> \frac{1}{2}$, see below). Second, since charged particles with spin carry intrinsic magnetic moments as well as charges, their behavior in a background magnetic field rather than a background electric field is easier to determine.

III. Asymptotic Freedom

In order to properly understand the sign difference which makes the gauge fields differ from the lower-spin fields, one must have a completely general discussion, valid for arbitrary-spin "elementary" particles. By elementary we shall mean here particles with a charge and a magnetic moment ($g = 2$) but no higher electromagnetic moments.

The perturbative vacuum of charged fields is a medium of particles with one particle present in each positive energy mode of the classical field in the case of Bose fields, and one particle in each negative energy mode in the case of Fermi fields [6]. In both cases the uniform charge density may be ignored; the correlations in the charge and current density fluctuations produce the permeability effects. With the aid of the uniform background magnetic field, as was first pointed out by N. K. Nielsen [7], one can derive the following simple expression, valid for arbitrary spin, for the magnetic susceptibility (and thus for $-\psi(e^2)/e^2 = -\psi_0 e^2$):[1]

$$-\psi_0 = \frac{1}{16\pi^2}\mathrm{Tr}\left[(2S_z)^2 - \tfrac{1}{3}\right] \times \begin{pmatrix}+1, \text{ bosons} \\ -1, \text{ fermions}\end{pmatrix}. \tag{3.1}$$

In the appendix there is a brief outline of an elementary derivation of (3.1).

The physical origin of the $-\frac{1}{3}$ term in (3.1) is the Landau diamagnetism, which is a consequence of the quantization of the orbital currents of the vacuum particles in the background magnetic field. The spin-dependent term is produced by the Pauli paramagnetism associated with the intrinsic magnetic dipole carried by charged particles with a spin. Although the result, $-\chi(\text{Landau}) = \frac{1}{3}\chi(\text{Pauli})$ for free electrons, is derived for non-

---

1. I would like to give this formula a name, the Nielsen-Hughes formula for ψ_0. This formula is valid for charged particles with *any* spin as long as the g-value associated with the intrinsic magnetic moment is equal to *two*, and so long as the particles carry no higher intrinsic moments.

relativistic particles [10], it continues to be true for relativistic elementary ($g = 2$) particles, generalized to arbitrary spin as it is in (3.1).[2]

The numerically "large" paramagnetism (and thus antiscreening) of elementary vector particles is a consequence of the g-value of spin magnetism being equal to 2 as opposed to a g-value of 1 for orbital magnetism and to the presence of the factor of $\frac{1}{3}$ in the diamagnetic term.

It should be stressed that, viewed in this way, asymptotic freedom at least to leading order does not test the Yang-Mills or non-Abelian character of the spin one field theory. Abelian gauge invariance requires the diagram with magnitude e^2, but to leading order, the meson-meson vertex is absent in the determination of ψ_0. Of course, renormalizability to higher orders requires that diagram with the same coupling constant e^2 be present. This is what makes the charged massless vector meson theory the same as the Yang-Mills gauge theory. It should also be noted that the formula (3.1) applies for higher spin particles with $g = 2$ [6], whether or not the resulting field theories turn out to be renormalizable [11].

The restriction to $g = 2$ is presumably a simple consequence of relativistic kinematics. For example, the motion of the "spin" associated with a relativistic particle in the presence of a magnetic field B is given by the formula [12]

$$\frac{d\mathbf{S}}{dt} = \left(\frac{e}{2m}(g - 2) + \frac{2e}{2E}\right)\mathbf{S} \times \mathbf{B} + \frac{e}{2m}(g - 2)\frac{E}{E + m}(\mathbf{v} \cdot \mathbf{B})\mathbf{v} \times \mathbf{S}. \qquad (3.2)$$

Only for particles where $g = 2$ is the coupling of energetic ($E \gg m$) particles independent of an explicit dependence on the mass (or alternatively, independent of a new scale associated with an "anomalous" moment, $\mu = (g - 2)e/(2m)$. This is a necessary condition for the absence of a quadratic divergence in the susceptibility in lowest order. We see from (3.2) that for massless particles, a "magnetic moment" with $g = 2$ can be associated with the *energy* of the particle. Field theories of higher-spin charged particles are in general not renormalizable but they can still satisfy this requirement, which makes the second-order charge re-

---

2. From (3.1), one can see that the familiar diamagnetism of the vacuum charges (and thus screening) for spin-0 bosons and spin-$\frac{1}{2}$ fermions have rather a different physical origin. Electrons "would be" paramagnetic because of their spin but for the fact that they carry negative energy in the vacuum state [6].

normalization diverge only logarithmically rather than quadratically. Of course, one might give a more explicitly quantum-mechanical derivation of the necessity of $g = 2$ for arbitrary spin by showing that the general form of the vertex graph ⌇∧ for massive charged particles is smooth in the limit as the mass vanishes only if g is taken as 2. In this case, the presence of a "magnetic" vertex when $m = 0$ requires the explicit introduction of an additional intrinsic scale μ for the magnetic moment just as in the classical theory. In (3.2) we may put $m = 0$, if $(g - 2)e/(2m)$ is first replaced by μ. Of course, there can be no real difference between the classical and quantum arguments since the correspondence principle is exact to the linear approximation in the external field.

When $g = 2$, we can see from correspondence principle that (3.2) is equivalent to

$$[\mathbf{S}, \omega^2] = i2e\mathbf{S} \times \mathbf{B} \tag{3.3}$$

or,

$$\omega^2 = (p - eA)^2 - 2e\mathbf{S} \cdot \mathbf{B}, \tag{3.4}$$

where $(p - eA)^2$ is the spin-independent part of the single-particle Hamiltonian and ω^2 gives the single-particle frequencies in the background field. (3.4) is the starting point of a derivation of the Nielsen-Hughes expression (3.1). Indeed, if the susceptibility is computed using (3.4), one obtains (3.1) (see appendix). One possible problem for particles with spin greater than one-half should be noted. In a uniform background field along the z-axis, the eigenvalues of $(p - eA)^2$ are the Landau levels $(2n + 1)eB + p_z^2$ with $n = 0, 1, \ldots$. As a consequence, if $s_z > 1$, there are a class of eigenvalues of (3.4) where $\omega^2 < 0$; these are the Nielsen-Olesen unstable modes [13]. The presence of these unstable modes makes the vacuum energy in the background field complex when the spin is greater than $\frac{1}{2}$. This complication has been ignored here.

The following historical note might be interesting to some. Asymptotic freedom or $\psi_0 < 0$ has been shown to be a consequence of the paramagnetism of charged vector particles. The first calculation of the charge renormalization of vector particles was made in 1964 [14], and although incorrect in magnitude, it was correct in sign. The result obtained was equivalent to $\psi_0 = \dfrac{-1}{16\pi^2}\left(\dfrac{20}{3}\right)$ rather than $\psi_0 = \dfrac{-1}{16\pi^2}\left(\dfrac{22}{3}\right)$, as one should obtain using (3.1). Now $\frac{22}{3} = \frac{20}{3} + \frac{2}{3}$, and if one looks closely at the earlier calculation one may surmise that there was but one error: the computation was made in a covariant formalism, but there was no inclusion of the necessary scalar *ghost* loops [15] present in such a formalism. These would give an additional "Landau" contribution $\left[2 \times \left(-\frac{1}{3} \right) \times (-1) \right] \times \left(\frac{-1}{16\pi^2} \right)$ to ψ_0. However, even if the calculation had been done correctly, it probably would have made no difference in 1964. After all, the sign of ψ_0 was obtained correctly and it was "unphysical" according to the authors of this work. These workers did not realize that they were computing the Gell-Mann and Low function of a Yang-Mills theory, and therefore that of a renormalizable field theory. Their suspicion was that the "unphysical" result of antiscreening was perhaps a consequence of the nonrenormalizability of a charged vector meson field theory [14]. Of course, the "Higgs" mechanism [16] to give masses to gauge bosons was just being formulated in the context of Abelian gauge field models, and it would take even more time until it was used to develop the present unified theory of weak and electromagnetic interaction [17] in the context of a non-Abelian theory. Furthermore, no one was thinking realistically about the strong interactions in the context of any field theory, much less a gauge theory, in 1964.

IV. Speculation about the "True" Vacuum of QCD

Asymptotic freedom has been shown to be a consequence of the paramagnetic properties of the "zero point" vector particles present in the perturbative vacuum state. As a consequence of the corresponding increase of the effective coupling over long-distance scales, one might anticipate a change in the form of the vacuum wave functional for the amplitudes associated with long-wavelength modes which would be such to suppress the infrared divergences associated with the perturbative running coupling constant. It shall be assumed [9] that it is still possible to approximately characterize the true ground state by a set of quasiparticle amplitudes associated with an expansion of the gluon field. That is, the vacuum state will be described as containing a certain number of "real" quasi-particles, together with "zero point" quasi-particles whose long-scale wavefunctions are highly modified, but whose short-scale wavefunctions are still plane waves. The later feature is, of course, a consequence of asymptotic freedom.

To produce such a description, let all space be subdivided into a lattice of cells of size R (linear dimension). In this way an explicit scale is introduced. The gluon field $A^k(x)$ (in the gauge $A^0 = 0$) can then be expanded by using a complete set of wavefunctions which do not appreciably overlap between cells. The quasi-particles are to be associated with the amplitudes of this expansion. The minimum wavelength of a quasi-particle will be of order $1/R$. An ansatz for a trial ground state can be given in terms of the quasi-particles defined by such an expansion of the field. It is proposed that it will contain some "real" quasi-particles in each cell along with the zero-point particles. The expectation of the energy density T^{00} of the gluon field may now be calculated perturbatively using the scale R. To zeroth order, $\langle T^{00} \rangle'$ in the state will be higher than for the perturbative vacuum defined by a plane-wave expansion of the field, so that

$$\langle T^{00} \rangle' = \frac{A}{R^4},$$

where the zero of energy has been defined by the energy of the perturbative vacuum. Clearly, $A > 0$. The energy density $\langle T^{00} \rangle'$ can now be computed to higher orders in α_S perturbatively, but with the state which is characterized by the scale R. The procedure of Gell-Mann and Low shall be implemented. A coupling $\alpha_S(R)$ which runs is introduced, that is, an $\alpha_S(R)$ which obeys the equation

$$-R^2 \frac{\partial}{\partial R^2}(\alpha_S(R)) = \psi(\alpha_S(R)).$$

Because the stress tensor obeys the equation $\partial_\mu T^{\mu\nu}(x) = 0$, it does not require a multiplicative renormalization, and so the only place that ultraviolet divergences can appear in $\langle T^{00} \rangle$ is in an additive constant which can be removed by a subtraction; so in terms of the coupling parameter $\alpha_S(R)$, we have

$$\langle T^{00} \rangle = \frac{1}{R^4}(A - \alpha_S(R)B). \tag{4.1}$$

Instead of defining $\psi(\alpha)$ in terms of the magnetic susceptibility, as was done in the earlier case of the electrodynamics of massless charged vector mesons, it can be defined directly in terms of the energy density in this calculation. As is well known, changes in the definition of the running charge $\alpha(R)$ do not affect the values of ψ_0 and ψ_1, but appear first at the level of ψ_2.[3] We may define ψ so that there are no higher-order terms in (4.1).

Because of the attractive perturbative force between gluons, if the cells are filled with gluons with total color and spin equal to zero, we might hope that B may be made large and positive, without at the same time making A too large. If with the best set of gluon wave functions this can be achieved, then one can see, as a consequence of "asymptotic freedom" (that is, $R\partial\alpha_S/\partial R > 0$), that $\langle T^{00}\rangle'$ will become negative and achieve a minimum for a value of R which is only slightly larger than that which makes $\langle T^{00}\rangle'$ become negative, i.e., $\alpha_S \approx (A/B)$. If the minimum occurs in a region where α_S is not too large then the state $|0\rangle'$ will give an approximate "vacuum" with energy lower that the perturbative state but presumably larger than the "true" vacuum. If α_S is small it will be in the domain where ψ is unique. Clearly, this means that it must be possible to choose the wave function and state so A is as small as possible, B as large as possible. The value of R which minimizes $\langle T^{00}\rangle'$ will be expressed in units of our basic scale R_0 (or Λ). Clearly if this is chosen so $\alpha_S(R_0) = 1$, then $R_{\min} \approx R_0$ if $A/B \approx 1$. This would also mean that $\langle T^{00}\rangle'$, although negative, would not be numerically large in units of R_0.

At present, although some estimates of A and B have been made on the basis of these ideas, the work has not gone far enough to be sure whether or not they might be accurate enough. The most important consequence of such a simplified ground-state wave function would be its ability to allow one to compute the low-lying excited states of QCD, that is, the "gluballs" spectrum. It is here that this conjectured state will find its real test, its ability to withstand the confrontation of experimental fact.

Appendix

The vacuum energy for charged particles is given by $\pm \mathrm{Tr}(\omega)$, $+$ for bosons and $-$ for fermions, where ω^2 is the operator

$$\omega^2 = (p - eA)^2 - 2eB \cdot S_z.$$

Here A is the potential which describes the uniform background field.

3. This can be simply demonstrated by making the change $\alpha'(R) = f(\alpha(R)) = \alpha(R) + \alpha^2(R)f_1 + \cdots$ and definition $- R\partial/\partial R\alpha' = \tilde{\psi}(\alpha')$ for $\tilde{\psi}$.

S_z is the spin operator for the massless particle. The operator $\omega_L^2 = (p - eA)^2 = p_z^2 + v_x^2 + v_y^2$, with $[v_x, v_y] = ieB$ has the Landau spectrum $p_z^2 + (2n + 1)eB$, with $n = 0, 1, \ldots$. The Landau "diamagnetism" can be derived from $\mathrm{Tr}(\omega_L)$ in an elementary way [10]. We quote the result for the energy per unit volume,

$$\frac{E^{\text{Landau}}}{V} = \frac{1}{24} e^2 B^2 \int_{\sim\sqrt{B}}^{\Omega} \frac{d^3 p}{|p|^3 (2\pi)^3} \begin{pmatrix} +1, \text{ bosons} \\ -1, \text{ fermions} \end{pmatrix} \mathrm{Tr}(1). \tag{A.1}$$

The trace is taken over spin states. The divergent part of the spin-dependent term can be gotten in an elementary way by simply expanding

$$\omega = (\omega_L^2 - 2eS_z B)^{1/2}$$

$$\approx \omega_L - \frac{eS_z B}{\omega_L} - \frac{1}{8}(2eS_z)^2 B^2 \frac{1}{\omega_L^3} + \cdots,$$

which then gives

$$\frac{E^{\text{Pauli}}}{V} = -\mathrm{Tr}\left[(2S_z)^2\right] \frac{B^2}{8} e^2 \int_{\sim\sqrt{B}}^{\Omega} \frac{d^3 p}{(2\pi)^3} \frac{1}{|p|^3} \begin{pmatrix} +1, \text{ bosons} \\ -1, \text{ fermions} \end{pmatrix}. \tag{A.2}$$

With $E/V \approx -\frac{1}{2}\chi B^2$, where χ is the magnetic susceptibility, and a combination of (A.1) and (A.2), one obtains the result (3.1).

For completeness, the well-known rule for obtaining ψ_0 for a gauge theory based upon $SU(n)$ from the above computation will be given. The $SU(2)$ charge equivalent to $SU(n)$ for the gauge particles is $e^2 = g^2 n/2$. The particle charge for particles belonging to the defining representation of $SU(n)$ is independent of n and equal to $g^2/2$ $[= (g/2)^2 + (-g/2)^2]$. For particles which belong to the adjoint representation (which occur, for example, in reference [11]) the charge is the same as for the gauge particles. Using the rules for $SU(3)$ and N_F triplet fermions, we obtain for ψ_0, $-\frac{1}{16\pi^2} \cdot \left\{ \frac{11}{3} \cdot \frac{3}{2} - N_F \cdot \frac{2}{3} \cdot \frac{1}{2} \right\} \times 2$ (the factor 2 for the two spin states), so the famous result, $-\frac{1}{48\pi^2}(33 - 2N_F)$ is obtained.

Acknowledgment

This work was supported in part through funds provided by the U.S. Department of Energy under contract DE-ACO2-76ERO3069.

References

[1] M. Gell-Mann and F. E. Low, *Phys. Rev.* 95: 1300 (1954). See also K. G. Wilson, *Phys. Rev.* 179: 1499 (1969); K. Symanzik, *Comm. Math. Phys.* 18: 227 (1970); C. G. Callan, *Phys. Rev. D* 2: 1541 (1970).

[2] E. D. Bloom et al., *Phys. Rev. Lett.* 23: 930 (1969); M. Briedenbach et al., *ibid.*, 935.

[3] R. P. Feynman, *Phys. Rev. Lett.* 23: 1415 (1969); J. D. Bjorken, E. Paschos, *Phys. Rev.* 185: 1975 (1969); S. D. Drell, D. J. Levy, T. M. Yan, *Phys. Rev.* 187: 2159 (1969); J. Kuti, V. F. Weisskopf, *Phys. Rev. D* 4: 3418 (1971).

[4] D. J. Gross and F. Wilczek, *Phys. Rev. Lett.* 30: 1343 (1973); H. D. Politzer, *ibid.*, 1346. I. B. Khriplovitch, *Sov. J. Nucl. Phys.* 10: 235 (1970); G. t' Hooft, unpublished; V. S. Vanyashin, M. V. Terentev, *JETP (Sov. Phys.)* 21: 375 (1965).

[5] T. Appelquist, M. Dine, I. F. Muzinich, *Phys. Lett. B* 69: 231 (1977); F. Feinberg, *Phys. Rev. Lett.* 39: 316 (1977); W. Fischler, *Nucl. Phys. B* 129: 157 (1977); V. N. Gribov, SLAC-TRANS-176 (1978) (12th Winter School, Leningrad Nuc. Phys. Inst., 1977); S. D. Drell, *New York Acad. of Sci. Trans. 11* 40: 76 (1980).

[6] G. R. Allcock, *Acta Phys. Pol. B* 11: 875 (1980).

[7] N. K. Nielsen, Odense preprint (1980). This work has since been published in *Am. J. Phys.* 49: 1171 (1981).

[8] R. J. Hughes, *Phys. Lett. B* 97: 246 (1980); R. J. Hughes, *Nucl. Phys. B* 186: 376 (1981).

[9] This is a brief summary of work in progress in collaboration with T. H. Hansson and C. Peterson.

[10] See, e.g., F. Seitz, *Modern Theory of Solids*, McGraw Hill, New York and London (1940) pp. 583–587.

[11] For an application to higher-spin theories, see T. L. Curtright, *Phys. Lett. B* 102: 17 (1981).

[12] V. Bargman, L. Michel, V. L. Telegdi, *Phys. Rev. Lett.* 2: 435 (1959).

[13] N. K. Nielsen, P. Olesen, *Nucl. Phys. B* 144: 376 (1978).

[14] V. S. Vanyashin, M. V. Terentev, *JETP (Sov. Phys.)* 21: 375 (1965).

[15] R. P. Feynman, *Acta Phys. Polonica* 24: 697 (1963).

[16] P. W. Higgs, *Phys. Rev. Lett.* 12: 132 (1964); P. W. Higgs, *Phys. Rev. Lett.* 13: 508 (1964); F. Englert, R. Brout, *Phys. Rev. Lett.* 13: 321 (1964); G. S. Guralnik, C. R. Hagen, T. W. Kibble, *Phys. Rev. Lett.* 13: 585 (1964).

[17] S. L. Glashow, *Nucl. Phys.* 10: 107 (1959); S. Weinberg, *Phys. Rev. Lett.* 19: 1264 (1967); A. Salam, in *Elementary Particle Theory, Relativistic Groups and Analyticity*, edited by N. Svartholm, Almquist and Wiksell, Stockholm (1964); S. L. Glashow, J. Iliopoulos, L. Maiani, *Phys. Rev. D* 2: 1285 (1970); G. 't Hooft, *Nucl. Phys. B* 35: 167 (1971).

The Running Coupling Constant in Quantum Electrodynamics

Sidney D. Drell

Quantum electrodynamics (QED) and quantum chromodynamics (QCD) are both local renormalizable gauge theories, but they describe physical systems with very different properties. In QED the effective charge, or interaction strength, increases at *short* distances. In contrast the interaction strength, or color charge, in QCD grows at *large* distances, leading to color confinement, but weakens at *short* distances if the particle content is suitably restricted (i.e., if the number of fermion flavors $n_f \leq 16$). This behavior of QCD is known as asymptotic freedom and is frequently summarized in terms of the running coupling constant $\alpha_s(q^2)$, which vanishes logarithmically with increasing q^2.

The idea of the running coupling constant emerges from applying the renormalization-group ideas and techniques pioneered independently by Stueckelberg and Peterman [1] and by Gell-Mann and Low [2]. Considerable experimental and calculational effort has been devoted during the past few years [3] to verifying the predicted behavior of $\alpha_s(q^2)$. In particular, when testing QCD care is required to separate the short-distance aspects of a process which can be treated perturbatively in terms of quark and gluon constituents from the large-distance features of hadronization which must be described nonperturbatively.

Life is generally easier in QED. The fine-structure constant is much smaller, there is no confinement, and the technical challenge of higher-order perturbation calculations is much less forbidding in an Abelian gauge theory. As long as the direct calculations prove tractable to the desired precision, little practical value is realized in QED by applying the renormalization-group techniques and describing its behavior in terms of $\alpha(q^2)$. We simply calculate to higher orders in powers of the fine-structure constant $\alpha \approx \frac{1}{137}$ as defined on the lepton mass shell for soft photons ($q^2 = 0$). There are, however, specific examples where renor-

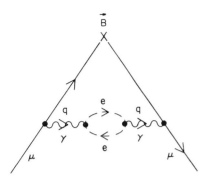

Figure 1

malization-group methods can simplify higher-order calculations. One such case is the higher-order vacuum polarization insertions in the calculation of the electron and muon magnetic moment anomalies, as shown in 1974 by Lautrup and de Rafael [4].

These vacuum polarization insertions give rise to a difference between the fourth-order magnetic moment anomalies of the electron and muon. In this contribution to the sixtieth birthday Festschrift for Francis Low, an admired colleague in physics, an intrepid partner in music, and a warm and close friend for many years, I will show how this well-known result of QED can be understood and reexpressed explicitly in terms of a running fine-structure constant $\alpha(q^2)$.

To order α^2, the electron and muon gyromagnetic anomalies, $\delta a \equiv \frac{1}{2}(g - 2)$, differ by the following amount:

$$\delta a_\mu - \delta a_e = \left(\frac{\alpha}{\pi}\right)^2 \left[\frac{1}{6}\ln\frac{m_\mu^2}{m_e^2} - \frac{25}{36} + O\left(\frac{m_e}{m_\mu}\right)\right],$$

$$\delta a_e = \frac{\alpha}{2\pi} - 0.328\left(\frac{\alpha}{\pi}\right)^2 + \cdots.$$

(1)

This difference arises from the contribution of the electron-positron bubble to the photon vacuum polarization in the $(g - 2)$ calculation for muons as shown in figure 1. The corresponding contribution due to a muon bubble in the electron calculation is reduced by powers of (m_e/m_μ), since it occurs only over a tiny distance scale.[1]

Consider first the calculation of the Schwinger term $\delta a \equiv \alpha/2\pi$, corre-

---

1. The electron bubble in the calculation of $(g - 2)_e$ and the muon bubble in $(g - 2)_\mu$ contribute identically.

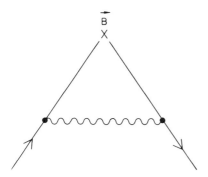

Figure 2

sponding to figure 2. To this order a bare photon is exchanged and its effective coupling to the electron or muon line is a *constant* α independent of the momentum running through the loop integral. When the electron bubble is included as shown in figure 1, the photon propagator is corrected to

$$
\begin{aligned}
\frac{-i}{q^2} \to \frac{-i}{q^2} &\left[1 - \frac{\alpha}{3\pi} \ln \frac{\Lambda^2}{m_e^2} + \frac{2\alpha}{\pi} \int_0^1 dz\, z(1-z) \ln \left\{ 1 - \frac{q^2 z(1-z)}{m_e^2 - i\varepsilon} \right\} \right] \\
&\approx \frac{-i}{q^2} Z_3 \left[1 + \frac{2\alpha}{\pi} \int_0^1 dz\, z(1-z) \ln \left\{ 1 - \frac{q^2 z(1-z)}{m_e^2 - i\varepsilon} \right\} \right].
\end{aligned}
\tag{2}
$$

Its effective charge now becomes

$$
\alpha \to \alpha(q^2) \equiv \alpha \left[1 + \frac{2\alpha}{\pi} \int_0^1 dz\, z(1-z) \ln \left\{ 1 - \frac{q^2 z(1-z)}{m_e^2 - i\varepsilon} \right\} \right],
\tag{3}
$$

with Z_3 being absorbed into the charge renormalization defined at $q^2 = 0$. Equation (3) defines the running coupling constant to order α. As appropriate for QED, which is not asymptotically free, α grows stronger with increasing q^2. In particular, α grows logarithmically[2] for large q^2:

$$
\alpha(q^2) \approx \alpha \left[1 + \frac{\alpha}{3\pi} \ln \frac{|q^2|}{m_e^2} \right], \quad |q^2| \gg m_e^2.
\tag{4}
$$

This growth of α accounts for the leading log in the difference[3] of $(g-2)$

---

2. When an infinite chain of such bubbles is included this contribution sums to a geometric series with the famous Landau ghost, i.e.,

$$
\alpha(q^2) = \frac{\alpha}{1 - (\alpha/3\pi) \ln (|q^2|/m_e^2)}.
$$

3. This is, of course, implicit in [4].

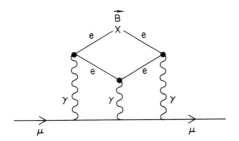

Figure 3

values in (1). In the calculation of the muon anomaly the mass of the muon sets the scale for the momentum running through the Feynman diagram in figure 1; i.e., $|q^2| \approx m_\mu^2$ in (4). In the calculation of the electron anomaly $|q^2| \approx m_e^2$, so that

$$\delta a_\mu - \delta a_e \approx \frac{\alpha}{2\pi} \cdot \frac{\alpha}{3\pi} \left[\ln \frac{m_\mu^2}{m_e^2} + O(1) \right], \tag{5}$$

in agreement with (1). It is apparent from (1) that the constant terms in (3) must be included in addition to the leading logarithm in order to reproduce the difference to any accuracy. From this comparison we learn that $|q^2| \approx (m_\mu/2)^2$ in the actual calculation of figure 1; i.e., roughly half the muon mass flows as momentum through the photon.

The fourth-order vacuum polarization insertion also gives rise to a difference between the sixth-order electron and muon moment anomalies. However the major contribution to this difference arises from the scattering of light-by-light insertion [5] shown in figure 3. The numerical coefficient of the amplitude corresponding to figure 3 is much larger than that of the vacuum polarization contribution, and its contribution to δa is larger in the ratio of 4.3 to 1, despite the fact that the latter is enhanced by $[\ln(m_\mu/m_e)]^2$ whereas the scattering of light-by-light insertion is enhanced only by one power of $\ln(m_\mu/m_e)$. This result warns us that higher-order amplitudes in QED, and presumably also QCD, contain a lot more physics than can be summarized in terms of a running coupling constant alone.

References

[1] E. C. G. Stueckelberg and A. Peterman, *Helv. Phys. Acta* 26: 499 (1953).
[2] M. Gell-Mann and F. E. Low, *Phys. Rev.* 95: 1300 (1954).

[3] See invited talk by R. Hollebeck (SLAC-PUB-2829) and summary talk by S. D. Drell (SLAC-PUB-2833) to the 1981 International Symposium on Lepton and Photon Interactions at High Energies, Bonn (August 24–29, 1981).

[4] B. Lautrup and E. de Rafael, *Nucl. Phys. B* 70: 317 (1974).

[5] S. J. Brodsky and T. Kinoshita, *Phys. Rev. D* 3: 356 (1971).

Functional Formulation of the Renormalization Group   Kerson Huang

1. Prologue

There was a time in physics when everything came up Regge poles. One day in 1963 Francis Low and I came across a paper claiming to have derived Regge behavior[1] in πN scattering by the "renormalization group" method. We had difficulty following the derivation, because Francis did not know what the renormalization group was (and I certainly didn't know), even though we gathered it had something to do with a paper by Gell-Mann and Low [1]. So we looked up the reference, a book by Bogoliubov and Shirkov [2], which started the exposition promisingly enough:

Essentially the same renormalization group, but from a different point of view, was utilized by Gell-Mann and Low to obtain concrete information on the asymptotic behavior of electrodynamic Green's functions for large values of the momenta. However, the investigation of Gell-Mann and Low was not carried out sufficiently carefully . . .

As we studied the book, however, we got thoroughly frustrated. Francis kept saying, "All they do is write $1 = 1$ in different forms. How can you get anything out of it?"

We finally figured out the right way to do it, which (of course) was to return to the method of Gell-Mann and Low. A crucial point missed by Bogoliubov and Shirkov, we realized, was the assertion that *the scattering amplitude with floating renormalization point approaches a finite limit as all masses tend to zero*. With that, we knew how to use the renormalization group to study scattering amplitudes [3]. We found that it places no

---

1. "Regge behavior" means that the high-energy limit of the scattering amplitude at fixed momentum transfer (and not fixed angle) behaves as a power of the center-of-mass energy, the power being a function of the momentum transfer alone.

restriction at all on πN scattering. It does place restrictions on $\pi\pi$ scattering, but only at fixed angle, and not at fixed momentum transfer. Thus the renormalization group has nothing to say about Regge behavior.

Nothing earth-shaking came of this, but we had fun. Francis threw all his papers in the waste basket as soon as he understood the point, while I kept a notebook.

"I hope you don't lose the notebook," Francis said. "Leave it to me in your will."

To spare Francis the wait, I have decided to reveal the secrets in that notebook, after almost twenty years.

2. Method of Gell-Mann and Low

Let us review what Gell-Mann and Low [1] did, from a point of view that will be useful for later applications.

The unrenormalized photon propagator in quantum electrodynamics has the form (in Landau gauge)

$$D'_{\mu\nu}(k) = \left(g_{\mu\nu} - \frac{k_\mu k_\nu}{k^2}\right) D'(k^2) + \text{(longitudinal terms)}, \tag{2.1}$$

where D' is gauge invariant. It also depends on the unrenormalized fine-structure constant α_0, the physical electron mass m,[2] and a cutoff momentum Λ. It is a sum of Feynman graphs with logarithmic skeletal divergences as $\Lambda \to \infty$. We put

$$D'(k^2) = -\frac{i}{k^2 + i\varepsilon} d'\left(\frac{k^2}{m^2}, \alpha_0\right), \tag{2.2}$$

where d' depends on k only through k^2/m^2, because it is dimensionless. The dependence on Λ^2/m^2 is left understood.[3] We can calculate d' in perturbation theory through the following relations:

$$\left[d'\left(\frac{k^2}{m^2}, \alpha_0\right)\right]^{-1} = 1 - \alpha_0 \Pi'(k^2), \tag{2.3}$$

$i\alpha_0(k^2 g^{\mu\nu} - k^\mu k^\nu)\Pi'(k^2) = \text{sum of all proper photon self-energy graphs.}$

The function $\Pi'(k^2)$ may be regarded as a functional of d'. (For simplicity, we ignore the functional dependence on the electron propagator. This

---

2. We assume mass renormalization has been carried out.
3. Functions that depend on the cutoff are generally denoted with a prime, as d'.

does change the results.) Skeletal logarithmic divergence means that this functional depends on Λ as a parameter, and diverges logarithmically as $\Lambda \to \infty$. Charge renormalization consists of identifying the physical fine-structure constant α as

$$\alpha = \alpha_0 d'(0, \alpha_0), \tag{2.5}$$

the experimental value of which is $\alpha/4\pi \approx 1/137$. We can define a "running" coupling constant α_μ by renormalizing at a floating renormalization point $k^2 = \mu^2$:

$$\alpha_\mu \equiv \alpha_0 d'\left(\frac{\mu^2}{m^2}, \alpha_0\right). \tag{2.6}$$

Then

$$\alpha_\mu/\alpha = d'\left(\frac{\mu^2}{m^2}, \alpha_0\right)\Big/ d'(0, \alpha_0). \tag{2.7}$$

Since $\Pi'(k^2)$, regarded as a functional of d', is only logarithmically divergent, one subtraction will render the functional finite. We define a finite functional (i.e., finite in the limit $\Lambda \to \infty$) Π by

$$\Pi\left[\frac{k^2}{\mu^2}, \frac{m^2}{\mu^2}; \alpha_0 d'\right] \equiv \Pi'(k^2) - \Pi'(\mu^2). \tag{2.8}$$

Using (2.3), (2.6), and (2.8), we obtain

$$\left[\alpha_0 d'\left(\frac{k^2}{m^2}, \alpha_0\right)\right]^{-1} = \alpha_\mu^{-1} - \Pi\left[\frac{k^2}{\mu^2}, \frac{m^2}{\mu^2}; \alpha_0 d'\right], \tag{2.9}$$

which is the Dyson equation. If we change the renormalization point μ, the change in Π must be canceled by the change in α_μ, since the left side is independent of μ. From (2.7) we see that α_μ changes by a multiplicative constant. Thus the operations of changing the renormalization point form a group, the "renormalization group." The question is, does the group property of α_μ have physical consequences?

Let us define a finite function d by the functional equation

$$\left[\alpha_\mu d\left(\frac{k^2}{\mu^2}, \frac{m^2}{\mu^2}, \alpha_\mu\right)\right]^{-1} = \alpha_\mu^{-1} - \Pi\left[\frac{k^2}{\mu^2}, \frac{m^2}{\mu^2}; \alpha_\mu d\right], \tag{2.10}$$

with

$$d\left(1, \frac{m^2}{\mu^2}, \alpha_\mu\right) = 1. \tag{2.11}$$

Then (2.9) can be rewritten as

$$\alpha_0 d'\left(\frac{k^2}{m^2}, \alpha_0\right) = \alpha_\mu d\left(\frac{k^2}{\mu^2}, \frac{m^2}{\mu^2}, \alpha_\mu\right). \tag{2.12}$$

Since this is true for all μ, the right side is a renormalization-group invariant. More explicitly,

$$\alpha_\nu d\left(\frac{k^2}{\nu^2}, \frac{m^2}{\nu^2}, \alpha_\nu\right) = \alpha_\mu d\left(\frac{k^2}{\mu^2}, \frac{m^2}{\mu^2}, \alpha_\mu\right). \tag{2.13}$$

The standard renormalized propagator d_c is defined by choosing the renormalization point to be at $\nu^2 = 0$:

$$\alpha d_c\left(\frac{k^2}{m^2}, \alpha\right) \equiv \lim_{\nu \to 0} \alpha_\nu d\left(\frac{k^2}{\nu^2}, \frac{m^2}{\nu^2}, \alpha_\nu\right). \tag{2.14}$$

Hence, for all μ,

$$\alpha d_c\left(\frac{k^2}{m^2}, \alpha\right) = \alpha_\mu d\left(\frac{k^2}{\mu^2}, \frac{m^2}{\mu^2}, \alpha_\mu\right). \tag{2.15}$$

By setting $k^2 = \mu^2$, we obtain the relation

$$\alpha_\mu = \alpha d_c\left(\frac{\mu^2}{m^2}, \alpha\right). \tag{2.16}$$

Using this, we can rewrite (2.15) as what appears to be a functional equation for d_c:

$$d_c\left(\frac{k^2}{m^2}, \alpha\right) = d_c\left(\frac{\mu^2}{m^2}, \alpha\right) d\left(\frac{k^2}{\mu^2}, \frac{m^2}{\mu^2}, \alpha d_c\left(\frac{\mu^2}{m^2}, \alpha\right)\right). \tag{2.17}$$

However, this is an identity by derivation, and places no restrictions on the functional form of d_c. Therefore, the multiplicative group property of α_μ has no physical consequence by itself.

We now supply information, namely, the zero-mass limit of d exists:

$$\lim_{m \to 0} d\left(\frac{k^2}{\mu^2}, \frac{m^2}{\mu^2}, \alpha_\mu\right) = f\left(\frac{k^2}{\mu^2}, \alpha_\mu\right). \tag{2.18}$$

This statement can be proven to order α_μ^2 in perturbation theory [4]; we assume, with [1], that it is true in general. Then, for $k^2 \gg m^2$ and $\mu^2 \gg m^2$, (2.17) reduces to an equation with content:

$$d_c\left(\frac{k^2}{m^2}, \alpha\right) = d_c\left(\frac{\mu^2}{m^2}, \alpha\right) f\left(\frac{k^2}{\mu^2}, \alpha d_c\left(\frac{\mu^2}{m^2}, \alpha\right)\right). \tag{2.19}$$

To find the most general solution, put

$$x \equiv k^2/m^2,$$
$$y \equiv \mu^2/m^2, \tag{2.20}$$
$$K(x, \alpha) \equiv \alpha d_c(x, \alpha).$$

Then

$$\frac{K(x, \alpha)}{K(y, \alpha)} = f\left(\frac{x}{y}, K(y, \alpha)\right). \tag{2.21}$$

Since this holds for all y, we choose a special value to facilitate its solution. For any given value of α, choose y such that

$$K(y, \alpha) = 1. \tag{2.22}$$

This can be solved in principle to give

$$y = 1/\phi(\alpha), \tag{2.23}$$

where ϕ is some function. Thus we have

$$K(x, \alpha) = f(x\phi(\alpha), 1) \equiv F(x\phi(\alpha)). \tag{2.24}$$

The most general form of d_c consistent with (2.19) is therefore[4]

$$\alpha d_c\left(\frac{k^2}{m^2}, \alpha\right) = F\left(\frac{k^2}{m^2}\phi(\alpha)\right) \quad (k^2 \gg m^2). \tag{2.25}$$

To obtain a differential equation for d_c, rewrite (2.19) as

$$\alpha d_c(x, \alpha) = \alpha d_c(y, \alpha)f\left(\frac{x}{y}, \alpha d_c(y, \alpha)\right). \tag{2.26}$$

Differentiating both sides with respect to x at fixed y, and then setting $x = y$, we obtain

$$x\frac{\partial}{\partial x}d_c(x, \alpha) = d_c(x, \alpha)\psi(\alpha d_c(x, \alpha)), \tag{2.27}$$

4. Francis Low recalled: "When we first got the functional equation [eq. (2.19)], Murray said, 'There's probably no solution.'

"The next day, I wrote down a solution on the blackboard:

$\alpha d_c(k^2/m^2, \alpha) = (k^2/m^2)\phi(\alpha).$

'But what's the most general solution?' Murray asked.

"So we consulted T. D. Lee, who was in the next office (at the University of Illinois at Urbana). T. D. came in and wrote $F(\quad)$ around my answer."

where

$$\psi(z) \equiv z[\partial f(\xi, z)/\partial \xi]_{\xi=1}. \tag{2.28}$$

The solution is

$$\ln x = \int_{F(\phi(\alpha))}^{\alpha d_c(x, \alpha)} \frac{dz}{\psi(z)}. \tag{2.29}$$

The function $\psi(z)$, known as the Gell-Mann–Low function, can be calculated in perturbation theory. The lowest-order result is

$$\psi(z) \xrightarrow[z \to 0]{} z^2/12\pi^2. \tag{2.30}$$

If we use this in (2.29) as if it were exact, we obtain the Landau approximation:

$$d_c(x, \alpha) = \left[1 - \frac{\alpha}{12\pi^2} \ln \frac{x}{x_0}\right]^{-1}, \tag{2.31}$$

which shows the infamous "ghost" pole at $x \approx e^{137}$.

If quantum electrodynamics is a finite theory, then α_0 must be finite, and

$$\alpha_0 = \lim_{k^2 \to \infty} \alpha d_c\left(\frac{k^2}{m^2}, \alpha\right). \tag{2.32}$$

By (2.29), this implies that $\psi(z)$ has a zero at some $z = z_0$. If this were a zero of finite order, one could, by integrating (2.29), conclude that $z_0 = \alpha_0$, thus reducing the problem to an eigenvalue problem for α_0, i.e., $\psi(\alpha_0) = 0$. However, it is now known $\psi(z)$ cannot have a zero of any finite order [5, 6]. Hence it remains an open question whether α_0 could have arbitrary values, if it were finite.

In the formulation of the renormalization group by Callan [7] and Symanzik [8], one begins with (2.12), suitably generalized to an n-point Green's function. Then one adopts a differential approach as in (2.27). The function $\psi(z)$ is now commonly designated as $\beta(z)$.

3. Pion-Pion Scattering

Consider isoscalar pions interacting through the self-interaction $\lambda_0 \phi^4$, where λ_0 is the unrenormalized coupling constant. Other interactions will be taken into account in the next section. The off-mass-shell $\pi\pi$ scattering amplitude will be denoted by $T'(p, \lambda_0)$, where p collectively denotes all the external 4-momenta:

$$p = \{p_1, p_2, p_3, p_4\}. \tag{3.1}$$

The pion propagator is denoted by

$$\Delta'(k, \lambda_0) = \frac{id'(k, \lambda_0)}{k^2 - m^2 + i\varepsilon}, \tag{3.2}$$

where m is the physical pion mass.[5] Actually, T' and d' also depend on m, and on a cutoff momentum Λ, and are divergent when $\Lambda \to \infty$.[6] Lorentz invariance and the fact that T' and d' are dimensionless imply that they depend only on invariant dimensionless combinations of p, k, m, and Λ. To avoid clutter, we shall not indicate this explicitly, unless it is important to do so.

By examining the Feynman graphs for T' and d', we deduce the functional relations (the Dyson equations)

$$T'(p, \lambda_0) = \lambda_0 + K'[p; T', d'],$$
$$[d'(k, \lambda_0)]^{-1} = 1 + \Pi'[k; T', d']. \tag{3.3}$$

The functionals K' and Π' are defined by connected Feynman graphs that have logarithmic skeletal divergences. This means that K' and Π' depend logarithmically on Λ as a parameter. From the structure of the Feynman graphs we can deduce the following scaling properties:[7]

$$K'[p; Z^2 T', Z^{-1} d'] = Z^2 K'[p; T', d'],$$
$$\Pi'[k; Z^2 T', Z^{-1} d'] = Z\Pi'[k; T', d']. \tag{3.4}$$

Thus, multiplicative changes in the functions T' and d' lead to multiplicative changes in the functionals K' and Π'.

Since K' and Π' are only logarithmically divergent, they can be rendered finite by one subtraction, which we perform at floating renormalization points a and b, respectively:

$$K\left[\frac{p}{a}, \frac{m}{a}; T', d'\right] \equiv K'[p; T', d'] - K'[a; T', d'],$$
$$\Pi\left[\frac{k}{b}, \frac{m}{b}; T', d'\right] \equiv \Pi'[k; T', d'] - \Pi'[b; T', d']. \tag{3.5}$$

5. See note 2.
6. See note 3.
7. The proof is as follows: In nth order (with $n \geq 2$, even), a graph in K' has $(4n - 4)/2$ internal lines, and a graph in Π' has $(4n - 2)/2$ internal lines, the external lines being all omitted. Upon the replacement $T' \to ZT'$, $d' \to Z^{-1}d'$, we have

$$K' \to Z^{2n-(2n-2)}K' = Z^2 K',$$
$$\Pi' \to Z^{2n-(2n-1)}\Pi' = Z\Pi'.$$

In the limit $\Lambda \to \infty$, K and Π approach finite functionals. We now rewrite the Dyson equations in the form

$$T'(p, \lambda_0) = T'(a, \lambda_0) + K\left[\frac{p}{a}, \frac{m}{a}; T', d'\right],$$

$$[d'(k, \lambda_0)]^{-1} = [d'(b, \lambda_0)]^{-1} + \Pi\left[\frac{k}{b}, \frac{m}{b}; T', d'\right].$$

(3.6)

Renormalization is carried out by assuming that there are finite functions T and d such that

$$T'(p, \lambda_0) = Z_b{}^2 T\left[\frac{p}{a}, \frac{m}{a}, \frac{b}{a}, \lambda_{ab}\right],$$

$$d'(k, \lambda_0) = Z_b{}^{-1} d\left[\frac{k}{b}, \frac{m}{b}, \frac{a}{b}, \lambda_{ab}\right].$$

(3.7)

Upon substitution into (3.6), we find the definitions

$$\lambda_{ab} = Z_b{}^{-2} T'(a, \lambda_0),$$

$$Z_b = [d'(b, \lambda_0)]^{-1}.$$

(3.8)

The second equation implies

$$d\left(1, \frac{m}{b}, \frac{a}{b}, \lambda_{ab}\right) = 1.$$

(3.9)

Combining the two equations in (3.8) gives

$$\lambda_{ab} = T'(a, \lambda_0)[d'(b, \lambda_0)]^2,$$

(3.10)

which defines the running coupling constant. By (3.7), it can also be expressed as

$$\lambda_{ab} = T\left(1, \frac{m}{a}, \frac{b}{a}, \lambda_{ab}\right).$$

(3.11)

Equation (3.7) implies that Td^2 is a renormalization-group invariant, for it is independent of the renormalization point. Let us define standard renormalized quantities T_c and d_c (the "physical" ones) by choosing $a = b = m$:

$$T_c(p, \lambda) \equiv T\left(\frac{p}{m}, 1, 1, \lambda\right),$$

$$d_c(k, \lambda) \equiv d\left(\frac{k}{m}, 1, 1, \lambda\right),$$

(3.12)

where $\lambda \equiv \lambda_{mm}$. Then, the invariance of Td^2 may be expressed as

$$T_c(p, \lambda)d_c^{\,2}(k, \lambda) = T\left(\frac{p}{a}, \frac{m}{a}, \frac{b}{a}, \lambda_{ab}\right)d^2\left(\frac{k}{b}, \frac{m}{b}, \frac{b}{a}, \lambda_{ab}\right).$$ (3.13)

Dividing this equation by one in which $k = b$, and using (3.9), we obtain

$$\frac{d_c(k, \lambda)}{d_c(b, \lambda)} = d\left(\frac{k}{b}, \frac{m}{b}, \frac{a}{b}, \lambda_{ab}\right).$$ (3.14)

By (3.11), (3.9), and (3.13), we can write

$$\lambda_{ab} = T_c(a, \lambda)d_c^{\,2}(b, \lambda).$$ (3.15)

Substituting this into (3.13) and (3.14), we obtain what seem at first sight to be functional equations for T_c and d_c:

$$T_c(p, \lambda)d_c^{\,2}(k, \lambda) = T\left(\frac{p}{a}, \frac{m}{a}, \frac{b}{a}, T_c(a, \lambda)d_c^{\,2}(b, \lambda)\right),$$ (3.16)

$$\frac{d_c(k, \lambda)}{d_c(b, \lambda)} = d\left(\frac{k}{b}, \frac{m}{b}, \frac{a}{b}, T_c(a, \lambda)d_c^{\,2}(b, \lambda)\right).$$ (3.17)

But, by derivation, these are mere identities, and impose no restriction on T_c and d_c.

Now we make the assumption that the functions T and d approach finite limits as $m \to 0$:

$$\lim_{m \to 0} T\left(\frac{p}{a}, \frac{m}{a}, \frac{b}{a}, \lambda_{ab}\right)d^2\left(\frac{k}{b}, \frac{m}{b}, \frac{a}{b}, \lambda_{ab}\right) = f\left(\frac{p}{a}, \frac{k}{b}, \frac{a}{b}, \lambda_{ab}\right),$$ (3.18)

$$\lim_{m \to 0} d\left(\frac{k}{b}, \frac{m}{b}, \frac{a}{b}, \lambda_{ab}\right) = g\left(\frac{k}{b}, \frac{a}{b}, \lambda_{ab}\right).$$ (3.19)

Then, for p^2, k^2, a^2, b^2 all $\gg m^2$, (3.16) and (3.17) become

$$T_c(p, \lambda)d_c^{\,2}(k, \lambda) = f\left(\frac{p}{a}, \frac{k}{b}, \frac{a}{b}, T_c(a, \lambda)d_c^{\,2}(b, \lambda)\right),$$ (3.20)

$$\frac{d_c(k, \lambda)}{d_c(b, \lambda)} = g\left(\frac{k}{b}, \frac{a}{b}, T_c(a, \lambda)d_c^{\,2}(b, \lambda)\right).$$ (3.21)

These functional equations do impose restrictions on T_c and d_c.

Consider the renormalized scattering amplitude T_c on the mass shell, where $p_i^{\,2} = m^2$, ($i = 1, 2, 3, 4$). There are two independent kinematic variables, $s = (p_1 + p_2)^2$ and $t = (p_1 - p_3)^2$, with

$$\frac{t}{s} = -\left(1 - \frac{4m^2}{s}\right)\sin^2\frac{\theta}{2},$$ (3.22)

where θ is the center-of-mass scattering angle. The condition under which the functional equations are valid is s, t, a^2, $b^2 \gg m^2$. Accordingly, we consider the kinematic limit $s \to \infty$ at fixed θ. We shall change notation slightly, writting $T_c(s, \theta, \lambda)$ and $d_c(k^2, \lambda)$ in place of $T_c(p, \lambda)$ and $d_c(k, \lambda)$. Put $k^2 = s$, $a^2 = b^2 = s_0$, and denote the renormalization-group invariant by

$$P_\theta(s, \lambda) \equiv T_c(s, \theta, \lambda) d_c^2(s, \lambda). \tag{3.23}$$

Then (3.20) and (3.21) read

$$P_\theta(s, \lambda) = f_\theta\left(\frac{s}{s_0}, P_\theta(s_0, \lambda)\right), \tag{3.24}$$

$$\frac{d_c(s, \lambda)}{d_c(s_0, \lambda)} = g\left(\frac{s}{s_0}, P_\theta(s_0, \lambda)\right). \tag{3.25}$$

These hold for any value of s_0, which we can choose at our convenience. For any given value of λ, choose s_0 such that

$$P_\theta(s_0, \lambda) = 1. \tag{3.26}$$

This defines s_0 as some function of λ:

$$s_0 = 1/\phi(\lambda). \tag{3.27}$$

Then (3.24) is equivalent to the statement

$$P_\theta(s, \lambda) = F_\theta(s\phi(\lambda)), \tag{3.28}$$

where F_θ is some function whose form depends on θ. Substituting (3.26) and (3.27) into (3.25), we obtain

$$\frac{d_c(s, \lambda)}{d_c(\phi^{-1}(\lambda), \lambda)} = g(s\phi(\lambda), F_\theta(s\phi(\lambda))), \tag{3.29}$$

or

$$d_c(s, \lambda) = D(\lambda)G_\theta(s\phi(\lambda)). \tag{3.30}$$

Therefore we obtain as final result

$$T_c(s, \theta, \lambda) = \alpha(\lambda)H_\theta(s\phi(\lambda)). \tag{3.31}$$

The differential forms of the functional equations, and the analog of the Gell-Mann–Low function, can be found in [3], and will not be discussed here.

4. Pion-Nucleon Interactions

We now add to the pion interactions a renormalizable coupling to nucle-
ons, e.g., $f_0 \bar{\psi} \gamma_5 \phi \psi$. Two questions are of interest. First, "Is the renormalization group relevant to πN scattering?" Second, "How will the new
coupling change the results of the last section?"

The answer to the first question is "no." The reason is that the Feynman
graphs for the πN scattering amplitude have no skeletal divergences, and
hence its kinematic variables do not participate in the renormalization-
group transformations.

To answer the second question, we outline the new functional equations
and their solutions. There are two renormalization-group invariants:

$$P(p, \lambda, f) \equiv T_c(p, \lambda, f) d_c^{\ 2}(p, \lambda, f),$$
$$Q(p, \lambda, f) \equiv \Gamma_c(p, \lambda, f) s_c(p, \lambda, f) d_c^{\ 2}(p, \lambda, f),$$

(4.1)

where Γ_c is the renormalized πN vertex function, $(p - M)^{-1} s_c$ is the
renormalized nucleon propagator, p denotes all relevant momenta, and
λ and f are the two renormalized coupling constants. For all momenta
and renormation points μ much larger than both the pion mass and the
nucleon mass, P and Q satisfy functional equations that are generalizations of (3.24):

$$P(p, \lambda, f) = F_1\left(\frac{p}{\mu}, P(\mu, \lambda, f), Q(\mu, \lambda, f)\right),$$
$$Q(p, \lambda, f) = F_2\left(\frac{p}{\mu}, P(\mu, \lambda, f), Q(\mu, \lambda, f)\right).$$

(4.2)

To solve the first equation, choose μ such that

$$Q(\mu, \lambda, f) = 1.$$

(4.3)

Then

$$\mu = 1/\phi(\lambda, f),$$
$$P(\mu, \lambda, f) \equiv \psi(\lambda, f).$$

(4.4)

Hence

$$P(p, \lambda, f) = F_1(p\phi(\lambda, f), \psi(\lambda, f), 1),$$

(4.5)

or

$$P(p, \lambda, f) = F(p\phi(\lambda, f), \psi(\lambda, f)).$$

(4.6)

Similarly, one can show that Q has the same form. One can also show that the generalization of (3.30) is

$$d_c(p, \lambda, f) = D(\lambda, f)G(p\phi(\lambda, f), \psi(\lambda, f)).$$ (4.7)

Therefore (3.31) is generalized to

$$T_c(p, \lambda, f) = \alpha(\lambda, f)H(p\phi(\lambda, f), \psi(\lambda, f)).$$ (4.8)

We can see the general pattern: when more couplings are added, T_c will acquire extra dependences on functions ψ_1, ψ_2, ... of all the coupling constants.

Acknowledgment

This work was supported in part through funds provided by the U.S. Department of Energy under contract DE-ACO2–76ERO3069.

References

[1] M. Gell-Mann and F. E. Low, *Phys. Rev.* 95: 1300 (1954).

[2] N. N. Bogoliubov and D. V. Shirkov, *Introduction to the Theory of Quantized Fields*, Interscience, New York (1959), ch. 8.

[3] K. Huang and F. E. Low, *Zh. ETF* 46: 845 (1964). English translation in *JETP* (*Sov. Phys.*) 19: 579 (1964). There is a grievous misprint in eq. (21), namely, in the argument of the function $F_0(x, \Phi_0(g))$, the comma after x should be omitted. The full consequence of the renormalization-group method is that the comma should be absent.

[4] R. Jost and J. M. Luttinger, *Helv. Phys. Acta* 23: 201 (1950).

[5] S. L. Adler, *Phys. Rev. D* 5: 3021 (1973); K. Johnson and M. Baker, *Phys. Rev. D* 8: 1110 (1973).

[6] M. Baker and K. Johnson, *Physica A* 96: 120 (1979), argue from the triangle anomaly that $\psi(\alpha)$ cannot have a zero at all. See also S. L. Adler, C. G. Callan, D. J. Gross, and R. Jackiw, *Phys. Rev. D* 6 2982 (1972); N. Christ, *Phys. Rev. D* 4 946 (1973).

[7] C. G. Callan, *Phys. Rev. D* 2: 1541 (1970).

[8] K. Symanzik, *Comm. Math. Phys.* 18: 227 (1970).

The Topological Bootstrap

Geoffrey F. Chew

I. Introduction

Thirty years ago in the early 1950s while at the University of Illinois I enjoyed a fruitful period of close collaboration with Francis Low, and ideas uncovered in my work with Francis started me on a lifelong quest for a theory that in the late fifties came to be described by the term "bootstrap." The general bootstrap idea is that all aspects of nature are determined by the requirement that they be consistent. No aspect is arbitrary; no aspect is "fundamental"; the combination of all aspects is self-determining. Because of human limitations the bootstrap idea can never be pursued in its full sense. Compromise is unavoidable, certain assumptions must be accepted in any human contemplation of consistency; but since the early 1950s the starting assumptions have shifted. At that time it was difficult to do without the Newtonian-Cartesian idea of objects moving in a spacetime continuum. Quantum mechanics, even with its emphasis on the discrete, was formulated within a continuous spacetime. I have come to see this assumption of an underlying continuum as the root of the celebrated paradoxes surrounding quantum mechanics. I believe that spacetime—as we experience it through our classical sense of a continuous world made of real objects—should emerge from a discrete quantum world. In this view continuous spacetime is an approximation—like the continuous thermodynamic notions of temperature and entropy—useful only in environments of appropriately great complexity. I shall in this talk describe an approach to the bootstrap idea in which local spacetime is not a conceptual starting ingredient.

The starting idea is a notion of "elementary event," with a past and a future but not embedded in continuous spacetime. One does not know at the beginning the precise meaning of a physical event; bootstrap theory must generate its own physical interpretation. We shall see

Figure 1
A connected graph representing a cluster of correlated elementary events.

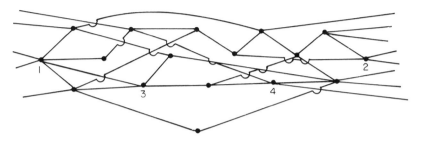

Figure 2
Graphs that illustrate the meaning of discrete interevent "distance."

that a physical event is a superposition of infinitely many patterns of correlated elementary events.

It is possible to make contact with two familiar notions by thinking of an elementary event as a *collision* between elementary particles or as an elementary-particle *decay*, both processes being "sudden"—that is to say, discrete. Mathematical meaning for the adjective "elementary" is provided by the concept of a *graph*. The term "elementary event" is to be understood as synonymous with a graph vertex and the term "elementary particle" is to be used synonymously with a graph line. The theory must generate a meaning for physical particles as well as for physical events.

The relation between correlated elementary events is assumed to be that between the vertices of a connected graph, as in figure 1. The vertices can be assigned a sequential order, but there is no metric—no meaning for distance between adjacent elementary events. A discrete notion of distance between nonadjacent elementary events is nevertheless achieved by counting the minimum number of vertices along any connecting path. For large distances such a measure becomes approximately continuous, as suggested by figure 2.

A postulate of topological bootstrap theory is that certain clusters of correlated elementary events are not to be distinguished from single elementary events. That is, the theory considers certain graphs as equivalent by a *contraction* process to a graph with a smaller number of vertices and lines. Strong interactions and strongly interacting particles (hadrons) will be distinguished from weak interactions by the contraction aspect of bootstrap theory. The rules for contraction relate to a notion of graph *complexity*, the idea being that contraction should not alter complexity.

I shall indicate to you how, by methods of combinatorial topology, one can associate a precise discrete measure of complexity to appropriately embellished event graphs. There are present at the same time consistency requirements on the definition of complexity, and out of these requirements flow constraints on elementary particles. Overall consistency of complexity-carrying event graphs not only controls the spectrum of elementary particles but determines physical masses and coupling constants. I shall sketch the so-far recognized consistency requirements and show how, in meeting these demands, there arise topological embellishments whose properties allow description by such terms as "quark," "gluon," "color," and "flavor"—terms occupying niches in the phenomenology of particle physics. Within the topological bootstrap such entities are not arbitrarily postulated; they emerge together with their properties as attributes of a consistent pattern of discrete causal connections. To immediately illustrate nonarbitrariness, but also to forestall misunderstanding, let me anticipate that topological quarks and gluons do not correspond to event-graph lines and thus do not emerge in the role of elementary particles; they are event-graph embellishments needed for consistency of the contraction principle.

The application of combinatorial topology to particle physics was initiated by Feynman in 1947 with his famous graphs whose lines carry energy and momentum—graphs introduced for photons and electrons but later recognized, especially by Landau,[1] as having more general significance. Feynman graphs constitute the event graphs of topological bootstrap theory even though not part of a rule for deducing the consequences of a local spacetime Lagrangian; there is no Lagrangian in the topological bootstrap.

Unembellished Feynman graphs are not adequate for dealing with *graph contraction*, an idea which emerged in 1969 through diagrams

---

1. I was pleased in 1959 to be told by Landau that he had been influenced by work of Francis Low and me on the connection between particles and the poles of analytic functions. Recognition of this connection was a major step in the evolution of bootstrap thinking.

invented independently by Harari and by Rosner [1]. Harari-Rosner diagrams when added to Feynman graphs allow a contraction rule— often characterized as "duality"—that is essential to the topological bootstrap. Harari-Rosner *and* Feynman diagrams are simultaneously indispensable to the theory I am describing. Also important are diagrams invented more recently by Finkelstein.

Bootstrap thinking in the late 1950s and early 1960s already recognized particles as "interevent connections" rather than as Newtonian-Cartesian "objects." Until contraction entered the game, however, there was no recognition that a *complexity hierarchy* is essential to consistency. Two decades ago bootstrappers spoke of a "nuclear democracy" in which all hadrons enjoy equal status. Today's topological bootstrap has uncovered a finite set of "elementary hadrons"—associated with a base level of topological complexity that we call "zero entropy." But even though an aristocracy, zero-entropy elementary hadrons are not arbitrarily assignable; they are determined by the demands of interevent consistency, and the contraction principle implies that each is a "composite"—built from other elementary hadrons.

II. Entropy and the Topological Expansion

In 1973 Veneziano [2] identified the notion of "topological entropy," so-called because this attribute of a complexity-carrying event graph cannot decrease when one graph is combined with another to form a new graph, as in the first step of figure 3. The second step illustrates how contraction represents the equivalence to a single event of certain *clusters* of causally connected elementary events. Uncontrolled cluster contraction evidently undermines meaning for "distance" within an event chain, but Veneziano (following Virasoro, Sakita, and others) realized that *embedding* the Feynman graph *in an oriented 2-dimensional surface*, which cyclically orders the lines incident on each vertex, allows a notion of complexity that controls contraction—as illustrated in figure 4, where an intermediate line remains unerasable. (Embedding graphs in manifolds of more than 2 dimensions is not useful for complexity theory.) Veneziano furthermore found that a suitably defined complexity never decreases under graph addition followed by contraction. The intermediate line in figure 4, for example, cannot be removed by adding further event graphs and then contracting. When *spin* as well as momentum is topologically represented it turns out, as I shall explain, that certain graph *vertices* as well

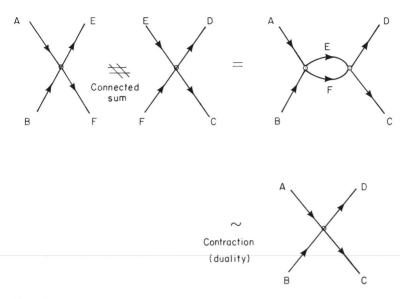

Figure 3
A connected sum of two single-vertex Feynman graphs that leads to a contractible
2-vertex graph.

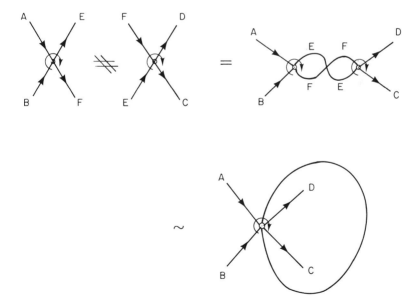

Figure 4
A connected sum of graphs with cyclically ordered vertices where an intermediate line
remains uncontractible.

as lines are unerasable. No vertex involving a photon, for example, can ever be removed by contraction.

The discovery of topological entropy through Feynman-graph 2-dimensional "thickening" opened the door to the notion of "topological expansion" for a physical-event amplitude—a complex number which according to quantum theory gives through its absolute value squared the probability for the event to occur. If M_{fi} designates the amplitude for a physical event where a collection of i ingoing particles leads to a collection f of outgoing particles, Veneziano's topological expansion reads

$$M_{fi} = \sum_{\gamma} M_{fi}^{\gamma},$$

where each value of the index γ—associated to a particular topological object—carries an entropy $g(\gamma)$. It is tacitly assumed that for some M_{fi} the expansion converges rapidly; i.e., that low entropy is more important than high entropy. Low energy emerges as essential to rapid convergence.

Where no contractions are possible, as turns out to be the case when hadrons are not involved, the idea of a topological expansion is essentially that of Feynman[2]—where the index γ refers to a Feynman diagram and the total number of graph vertices (minus 1) is an entropy index. Harari-Rosner contractions may change the number of vertices but they never alter the value of a legitimate entropy index. Finding such indices is the topological bootstrap game.

Implicit in the definition of entropy is that *zero-entropy* topological components cannot be built by addition from components with nonzero entropy. Most of the latter may be built from simpler (lower-entropy) components by a linear additive process, but zero-entropy components must be nonlinearly self-building. Herein lies the first bootstrap stage: The spectrum of *elementary hadrons* is postulated to be determined by zero-entropy consistency, and strong interactions are generally defined as those topological-expansion components generated by a "connected sum" of zero-entropy components. I shall explain later how, in a second stage, electroweak components enter the bootstrap.

Veneziano's topological expansion, formalized by Rosenzweig and me [3], is now called "classical DTU"—these initials standing for "dual topological unitarization."[3] Each value of γ corresponds to a single-

---

2. For elementary photons interacting with charged elementary leptons, a topological amplitude M_{fi}^{γ} has the same value as that given by the Feynman rules for quantum electrodynamics.

3. Classical DTU is restricted to mesons and ignores spin, but it nevertheless had impressive success—reviewed in [3]—in describing meson properties related to flavor and momentum.

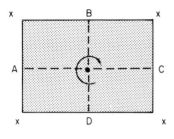

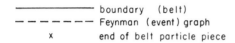

(a)

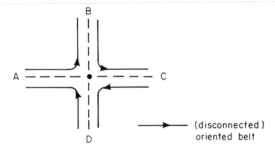

(b)

Figure 5
(a) Classical surface for a 4-meson zero-entropy event. (b) The corresponding Harari-Rosner diagram.

vertex Feynman graph embedded in a 2-dimensional oriented bounded "classical" surface whose boundary divides into "particle pieces" dual to the external lines of the graph. A zero-entropy topological component corresponds to a disk, such as illustrated in figure 5a. (The disk topology is usually called "planar.") "Addition" of two different components corresponds to an orientation-preserving connected sum along the boundary, those particle pieces of boundary being identified and erased that correspond to intermediate particles—as shown in figures 6 and 7.

In classical DTU two adjacent Feynman vertices contract to a single vertex and two "parallel" Feynman lines contract to a single line. There are two (nondecreasing) entropy indices: $g_1 = $ genus or twice the handle number and $g_2 = $ number of (noncontractible) closed loops, or

Chew   56

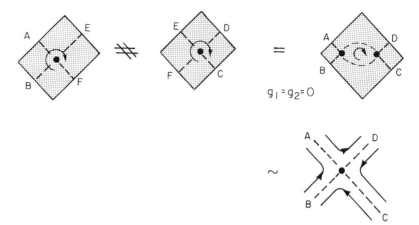

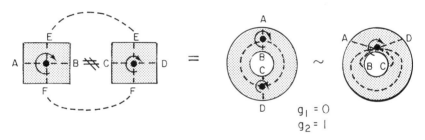

Figure 6
Connected-sum of two zero-entropy classical surfaces that leads to a zero-entropy surface
$(g_1 = g_2 = 0)$.

Figure 7
Connected sum of two classical surfaces that leads to a surface of nonzero entropy
$(g_1 = 0, g_2 = 1)$.

$$g_2 = b - 1 + g_1,$$

where b is the number of boundary components. As illustrated in figures 6 and 7, these two indices record the complexity of intermediate thickened Feynman-graph lines involved in building up a given topology starting from zero entropy ("planar" in classical DTU)—zero entropy being uniquely characterized by $g_1 = g_2 = 0$. The full (infinite) topological expansion contains components belonging to all possible values of g_1 and g_2.

The boundary of the classical surface inherits the surface orientation and constitutes a closed graph; if cut open at the Feynman-graph ends we have the globally oriented Harari-Rosner graph (see figure 5b). The

Harari-Rosner graph, that is to say, is the boundary of the oriented thickened Feynman graph.

The general rules for determining a topological amplitude M_{fi}^γ are based on S-matrix principles such as unitarity together with the graphical Landau prescription for the singularities of S-matrix connected parts [4]. Corrections to zero entropy turn out to be expressible through Feynman-like rules.

III. Spin and Electric Charge

In classical DTU the Feynman graph is recognized as carrying momentum and energy, and the topology γ of the classical embedding surface correlates with the Riemann surface that carries the momentum singularities of the amplitude M_{fi}^γ [3]. But no recognition is given in classical DTU to spin. Now momentum and spin are both "direct" classical observables; both emerge from the Poincaré group and both are needed to define an S matrix in a Hilbert space of asymptotic quantum states. To make contact with experiment topological particle theory requires an S matrix. Ultimately one hopes to explain as a manifestation of high complexity the notion of "measurement" which gives meaning to asymptotically observable momentum and spin, but presently the possibility of momentum and spin measurement is a starting assumption. Topological bootstrap theory, in exploring requirements of consistency within the S-matrix context, takes for granted both momentum and spin as particle characteristics. So where does spin reside on the classical surface?

As revealed by the work primarily of Mandelstam [5] and Stapp [6] but with clarification by Finkelstein [7], spin flows along the classical-surface boundary; that is, the Harari-Rosner graph is the carrier of spin in the sense that the spin-dependence rules for M_{fi}^γ are expressible through Harari-Rosner lines. Although I shall not here state the Mandelstam-Stapp rules, they employ the notion of chirality and are unambiguously determined by the consistency demands of zero entropy. The rules depend on the classical surface containing locally oriented patches that at zero entropy are defined by the Feynman (momentum) graph together with the boundary (Harari-Rosner) graph. Additional patch-boundary lines may be generated by mismatch of local orientations in connected sums, and constitute an ingredient of complexity which blocks contraction of vertices (see figure 8). I shall explain later that within strong-interaction topologies these added patch lines are appropriately called "topological vector gluons"; the number of such "gluons" is an entropy index g_3

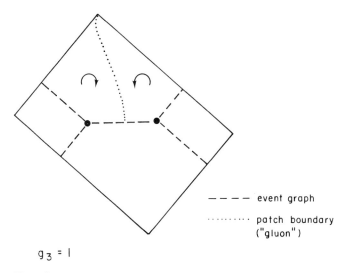

$g_3 = 1$

Figure 8
Example of classical surface with one unit of chiral complexity.

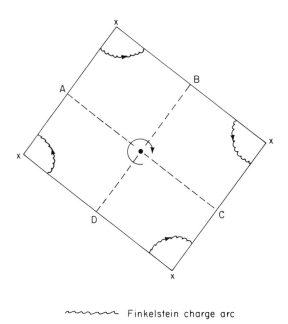

Figure 9
Example of charge lines on a zero-entropy classical surface.

which augments the g_1 and g_2 of Veneziano. I return later to further discussion of spin in connection with the quantum-mechanical notion of "fermion" but pass now to the final classical (asymptotic) observable—electric charge—whose measurement inevitably intertwines with that of momentum and spin.

One expects classical particle topology to contain some feature related to electric charge. An embellishment consistent with classical electromagnetism as well as with S-matrix requirements is a set of nonintersecting directed lines on the classical surface, each of which begins and ends on a different boundary subpiece without crossing a momentum line. Figure 9 provides an example by embellishing figure 5a with 4 charge arcs. With n charge lines the boundary then divides into $2n$ subpieces. Belonging to each particle is a boundary portion consisting of several subpieces, while on each subpiece lies the end of exactly one charge line. Electric charge is quantized and conserved if the charge carried by each subpiece is zero when charge and boundary directions disagree and ± 1 when there is agreement of direction. Charge lines are indispensable for electromagnetism, but they turn out to be needed already for strong interactions. Their principal architect has been Jerry Finkelstein, although the papers describing this embellishment have included other authors because of correlated additional topological features [8].

IV. The Quantum Surface

The major further feature is the *quantum surface*, originally proposed by me and developed in detail through a lengthy, still-continuing, collaboration with Poénaru [9]. The primary question to be answered through the quantum surface is how many boundary subpieces attach to each elementary particle. It was shown by Weissmann [10] from the zero-entropy contraction aspects of classical DTU that the conditions for zero-entropy joining of one boundary particle-piece end to the end of a different particle piece must be independent of other ends. Because a particle piece of boundary has at least 2 ends, there must be for each particle at zero entropy at least 2 subpieces[4]—each with its own attached charge line. The number of subpieces per particle may, however, be larger than 2. The ambiguity is resolved through a closed oriented "quantum" surface, *transverse* to the classical surface and "thickening" the boundary thereof.

---

4. This requirement guarantees "quark confinement" because, as we shall see, "quarks" correspond to certain boundary subpieces.

Thickening means that the boundary graph is embedded in the quantum surface.

The quantum surface divides into oriented particle areas, one for each ingoing or outgoing elementary particle in the event; each particle area houses the corresponding particle piece of the classical-surface boundary. Division of a particle boundary piece into subpieces corresponds to a division on the quantum surface of the particle area into subareas inside each of which there ends one charge line, and consistency conditions on particle subdivisions flow from the requirement that at zero entropy the total quantum surface closes into a sphere. Certain conserved "internal" quantum numbers are thereby implied, together with zero-entropy symmetries that become broken at higher levels of the topological expansion.

Connected sums along boundaries of classical surfaces are accompanied by quantum-surface connected sums in which corresponding particle areas are identified and erased in such fashion as to preserve surface orientation. The orientation of the classical surface boundary (the Harari-Rosner graph) is inherited from the quantum-surface orientation—the latter providing the distinction between ingoing and outgoing particles or between particles and antiparticles.

The search for a consistent pattern of zero-entropy particle areas on the quantum surface has been lengthy and laden with surprises. Numerous patterns have been proposed and subsequently discarded. The pattern now to be described, found by Poénaru and me, has been stable for more than one year and has survived many consistency tests. Although no uniqueness proof has been achieved, we are aware of no satisfactory alternative.

Our pattern divides the quantum surface into *triangles* of alternating orientation, each triangle being "mated" to exactly one other triangle of opposite orientation—mating being defined as a sharing of all 3 vertices. The mate of a triangle in one particle area at zero entropy always lies in another particle area. Particle areas are triangulated disks; each subpiece of the classical-surface boundary thickens into exactly one quantum triangle.

Only three forms of particle area on the quantum surface are allowed at zero entropy—those shown in figure 10, where two types of triangle occur. A "peripheral" triangle contributes two edges to the particle-disk perimeter while a "core" triangle contributes no edges, although all triangle vertices lie along the perimeter. Figure 11 shows how the classical-surface boundary (or "belt," for short) cuts all triangles and at zero entropy always enters and leaves particle disks at trivial vertices, which

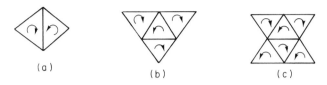

Figure 10
The allowable zero-entropy elementary-particle areas.

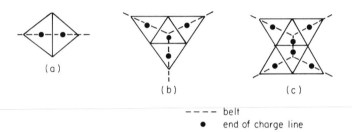

```
– – – –  belt
   •     end of charge line
```

Figure 11
Intersection of belt with zero-entropy particle areas. Ends of charge lines are also shown.

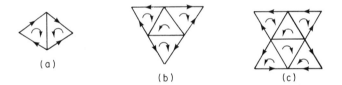

Figure 12
Edge orientations along perimeters of elementary-particle areas.

uniquely belong to peripheral triangles. Also shown in figure 11 are the ends of charge lines, one for each triangle. Finally, each edge along a particle perimeter is oriented, as shown in figure 12; these orientations must match when particle disks are fitted together on the quantum surface.

Looking at an individual fully embellished peripheral triangle, as in figure 13, we see a 2-fold electric-charge degree of freedom and a 4-fold edge-orientation degree of freedom, all of which must be matched at zero entropy by the mate of this peripheral triangle. These quantum numbers we associated with "flavor" and thus predict 8 flavors. The 4 edge flavors, usually called "generations," will be separately conserved on any orientable quantum surface with trivial vertices, although nonorientable sur-

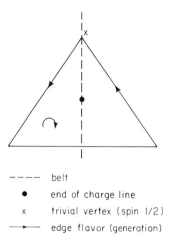

----   belt
●   end of charge line
x   trivial vertex (spin 1/2)
——►——   edge flavor (generation)

Figure 13
The peripheral triangle or "topological quark"; also called *I*-triangle.

faces allow generation mixing. Continuity of charge-line direction assures that electric charge is always conserved.

Each peripheral triangle not only mates with an "anti"–peripheral triangle but shares a trivial vertex therewith. The Mandelstam-Stapp rule for spin dependence at zero entropy centers on the quantum-triangulation trivial vertices that connect adjacent particle pieces on the classical-surface boundary. The rule effectively attaches spin $\frac{1}{2}$ to each quantum triangle sharing such a trivial vertex. Since the latter are always peripheral, fermion number is the number of clockwise peripheral triangles minus the anticlockwise number.

The full collection of peripheral-triangle attributes makes "topological quark" appropriate as a descriptive name for this type of triangle. It must be remembered, however, that topological quarks do not carry momentum; they are not elementary particles, even though they carry spin, electric charge, and flavor.

The quantum surface is not built entirely from topological quarks. Already at zero entropy there appear core triangles, which carry no spin or flavor but which are electrically charged and each of whose edges, correspondingly, can be said to carry one of 3 different "topological colors"—as shown in figure 14. Equivalently one may say 3 differently colored sheets of the classical surface meet at a "junction line" that ends inside a core triangle [9]. The sheet colored #1 carries the Feynman

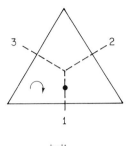

---- belt
• end of charge line

Figure 14
The core triangle; also called Y-triangle.

graph. Color may be attributed to a quark according to the classical sheet on which it lies and each topological color is separately conserved even though topological quarks of a given flavor may change color in connected sums that increase entropy. Each exchange of color between a pair of quarks builds a link between corresponding sheets of the classical surface, the number of such links being an entropy index g_4. Topological color, unlike QCD color, admits only discrete transformation (permutations) and color #1 is evidently not symmetric with respect to colors #2, 3.

The number of clockwise core triangles minus the anticlockwise number is conserved and, for strong-interaction topologies, may be identified with the negative of baryon number B. At this stage we are able to associate the disk of figure 10a with an "elementary meson"—quark plus antiquark with $B = 0$. The disk of figure 10b is an "elementary baryon"— 3 quarks and one anticore triangle, so $B = 1$. The disk of figure 10c is an "elementary baryonium"—2 quarks, 2 antiquarks, one core triangle, and one anticore triangle, so $B = 0$. The quantum numbers of low-mass mesons and baryons agree with topological bootstrap theory. The experimental failure to observe baryonium is discussed below.

Because all requirements for smooth joining of particle quantum disks must reside on disk perimeters (Weissmann) [10], the direction of a core charge arc is not variable. In fact zero-entropy consistency conditions require agreement between core charge and boundary (Harari-Rosner) orientations, so a clockwise (anticlockwise) hadron core triangle *must* have electric charge $Q = +1(-1)$, a point first noted by McMurray. Because topological quarks carry $Q = 1, 0$, the total electric charge of a hadron disk is

$$Q = N \, _{\text{quarks}}^{\text{charged}} - B.$$

At the same time, for any hadron (see figure 10),

$$B = \tfrac{1}{3} N_{\text{quarks}} = \tfrac{1}{3} (N \, _{\text{quarks}}^{\text{charged}} + N_{\text{quarks}}^{\text{neutral}}),$$

so

$$Q = \tfrac{2}{3} N \, _{\text{quarks}}^{\text{charged}} - \tfrac{1}{3} N \, _{\text{quarks}}^{\text{neutral}}.$$

Thus if hadron core triangles are ignored it appears that quarks carry charge $Q = \tfrac{2}{3}, -\tfrac{1}{3}$. For pure strong-interaction topologies a bookkeeping has been devised that makes no reference to core triangles—keeping track only of quarks—although this device does not work for electroweak interactions. With such bookkeeping the familiar fractional quark charges are appropriate.

I remark here that the complexity growth recorded by the two entropy indices g_3 and g_4 (now added to Veneziano's g_1 and g_2) can appropriately be described in quark-gluon language. One may speak of "vector gluons" emitted by topological quarks because, when a classical patch boundary line ends at a trivial vertex, the effect on quark spin is similar to that in perturbative field theory at a quark-gluon vertex. On the other hand, a color switch between two topological quarks is as if a "color-carrying gluon" were exchanged. Gluons in either sense are absent at zero entropy; conversely, large values of g_3 and g_4 correspond to large numbers of topological gluons.

V. Topological Supersymmetry

It was pointed out by Gauron, Nicolescu, and Ouvry [11] that zero entropy (where $g_1 = g_2 = g_3 = g_4 = 0$) is characterized by a "topological supersymmetry"—all elementary mesons, baryons, and baryoniums sharing a single (nonvanishing) mass m_0 even though the spin values $0, \tfrac{1}{2}, 1, \tfrac{3}{2}, 2$ occur. The zero-entropy S-matrix conditions allow a complete factorization of spin, flavor, color, and chirality degrees of freedom, so that only momentum remains in the nonlinear bootstrap equations representing unitarity. It is plausible that these equations admit no more than one solution and determine, among other things, a unique value for the zero-entropy 3-hadron dimensionless coupling constant g_0 [12]. (The mass m_0 simply sets the energy scale for the topological expansion.)

By means of a variety of approximations, Balazs, Finkelstein, and Espinosa [13] have attempted to calculate the zero-entropy coupling constant and have found that

$$g_0 \lesssim 2e.$$

The smallness of g_0 here derives from the large multiplicity f of closed elementary-hadron loops at zero entropy [12]. It turns out that $f = 32(31) \approx 10^3$ while $g_0^2/16\pi^2 \approx 1/f$. The number 32 is the total number of distinct zero-entropy topological quarks:

2 spins × 2 charges × 4 generations × 2 chiralities = 32.

Each closed momentum loop is accompanied by either 1 or 2 closed quark loops. Hence the net factor is $-32 + (-32)^2 = 32(31)$. The minus sign is that originally discovered by Feynman for any closed fermion loop and rediscovered by Stapp [6] as part of the topological spin rules. One consequence of the minus sign is that a consistent zero-entropy (hadron) spectrum cannot terminate with elementary mesons; elementary baryons and baryoniums must be included.

Gluonic (g_3, g_4) corrections to zero entropy break topological supersymmetry. A mechanism for breaking spin degeneracy is qualitatively understood, calculations by Levinson being underway to see whether such observed splittings as $\pi - \rho - N - \Delta$ can be semiquantitatively achieved through leading topological components with nonvanishing g_3. A tentative color-switch (g_4) mechanism for breaking generation symmetry has also been uncovered, based on topological-color asymmetry, but so far there is no visible strong-interaction mechanism to break charge-doubling symmetry (e.g., interchange of charmed and strange quarks). We anticipate that charge symmetry is first broken by electroweak interactions.

An argument can be made that dimensionless hadron coupling constants are less affected than hadron masses by corrections to zero entropy. Finkelstein, Levinson, and I [14] accordingly have calculated elementary 3-hadron physical coupling constants in the zero-entropy approximation where each is a simple multiple of g_0 and found that $SU(6)_W$ ratios emerge, as reported earlier by Mandelstam [5] on the basis of a less complete but compatible version of zero entropy. But topological supersymmetry goes further and gives the ratios of baryon-meson to meson-meson couplings. Thus *all* measured elementary-hadron coupling constants are explained by a single assignment of g_0. This value predicts enormous baryonium coupling constants and correspondingly large widths, so the experimental failure to find narrow baryonium states is understandable.

VI. Electroweak Interactions

From a bootstrap standpoint why should there be anything beyond hadrons and strong interactions? One reason for electromagnetism relates to the high-entropy limit corresponding to the classical world of real objects embedded in spacetime. I shall argue later that soft photons constitute the key to such a limit—which is needed for the measurements defining an S matrix. Topological representation of photons has been achieved through the ingredients already described [8] and brings along in a natural fashion 3 other electroweak vector bosons [15]. (A quartet of scalar bosons—charge doublet plus antidoublet—is also natural.) But why leptons?

A conjectured reason is motivated by standard Lagrangian perturbation theory which, at least for photon-lepton interactions, has the same content as a topological expansion. (A source of such equivalence is the fact that electroweak bosons have attached ingredients of topological complexity that block contraction.) Now Lagrangian electroweak perturbation theory has uncovered consistency problems which require quarks and leptons to be *paired*—one lepton matching each distinct quark but carrying opposite electric charge. We conjecture that topological theory will encounter a similar problem: A consistent interaction between electroweak vector bosons and quarks would then require quarks to be paired with leptons.

The present leading candidate for a lepton quantum-surface area is a (nonorientable) Möbius band built from one peripheral triangle and one neutral core triangle, with two edges of the latter identified. The spin and flavor content is the same as that of a quark, since the perimeter is built from the two oriented peripheral-triangle edges. There is, however, a momentum line ending inside the lepton area; so, unlike a topological quark, a lepton area corresponds to an elementary particle. Each electroweak boson corresponds to a closed quantum surface covered by 2 triangles. Vector-boson triangles are like that of figure 13 but without edge orientations.

Strong-interaction topology has been developing for more than 10 years and electroweak topology for less than 2; the structure of the former is correspondingly more secure. So far our electroweak topologies have been guided as much by QED and the Weinberg-Salam extension thereof as by consistency considerations. It is reassuring that the full content of QED is embeddable in topological particle theory, but we hope eventually to understand from consistency considerations not only the raison d'être for electroweak interactions, but also the zero photon mass together with other arbitrary aspects of Weinberg-Salam theory and eventually *CP* violation. It is possible that certain aspects of Weinberg-Salam theory will not be duplicated.

The breaking of *CP* invariance, as well as Cabibbo quark-generation mixing and proton decay, is related to *nonorientable* quantum and classical surfaces. We have a tentative understanding of why lepton generation mixing is weaker than Cabibbo quark mixing and why baryon-lepton transitions are *extremely* weak but perhaps not impossible.

VII. Spacetime

At this point of the story we are often asked about gravitons. But before that question we should first be asked about spacetime and the classical world of real objects embedded in spacetime. Our viewpoint is that objective reality and its accompanying spacetime acquire meaning through event patterns of high complexity. Gravitation, as an aspect of spacetime, will correspondingly be meaningful only in such a classical limit—where quantum effects have been smoothed away. We therefore do not anticipate significance for gravitons as elementary particles like photons.

It appears that soft-photon emission and absorption is the unique type of quantum event "gentle enough" to allow development of classical objectivity. Any photon event, no matter how soft, is noncontractible and contributes to growth of entropy; electromagnetism becomes recognized as the key to compatibility between the real world of classical measurement and the quantum world of elementary events. If so, "real" spacetime and gravitation are meaningful only in a photon-dominated environment.

What about the "short-distance parton" concept so valuable to the phenomenology of hadron events with large transfer of momentum? Here again as the key to significance we see high complexity, although not a complexity generated by gentle photons. Quite the contrary, here events are violent and large values for the classical-DTU entropy indices g_1 and g_2 become important. We must suppose that "partons at short distance" will acquire approximate meaning for sufficiently large average values of g_1 and g_2.

VIII. Accessible and Inaccessible Degrees of Freedom

Looking back over topological bootstrap theory, one distinguishes 3 categories of topological features. First there are the graphs invented by Feynman, Finkelstein, and Harari-Rosner that record, respectively, the flow of the direct observables: momentum, electric charge, and spin. Next there are the quantum-surface orientations that distinguish different types of elementary particle and that correspond to internal quantum numbers—revealed indirectly through event selection rules. The first category is directly accessible to experiment, and the second category is indirectly accessible. Finally there is a 2-dimensional classical surface which embeds both momentum and charge graphs and which has for boundary the spin graph. It is the intergraph relation through the classical surface that generates entropy from *inaccessible* features of the topology.

The momentum-entropy indices g_1 and g_2 emerge from the cyclic ordering of momentum lines incident on Feynman-graph vertices, an ordering without physical significance. The chirality-entropy index g_3 stems from oriented classical-surface patches defined by the momentum and spin graphs; the patch orientations are not accessible to experiment. The color-entropy index g_4 emerges from the location of Finkelstein's charge graph on the classical surface, a location also without physical meaning.

For a given event—specified by sets i and f of initial and final particles—the topological superposition building the event amplitude

$$M_{fi} = \sum_{\gamma} M_{fi}^{\gamma}$$

runs over inaccessible degrees of freedom, all accessible topological features being fixed by i and f. One may feel uneasy about the theory's dependence on physically unmeasurable variables, but it should be remembered that quantum mechanics, as formulated more than half a century ago, depends on an inaccessible but essential mathematical feature: the phase of a complex Hilbert-space vector. So topological bootstrap theory may be seen as an extension of quantum theory's complex numbers—to a broader domain of mathematical structures inaccessible to objective measurement but essential to overall consistency. Objective measurement promises itself eventually to emerge in a larger bootstrap as a necessary component of a consistent nature.

Acknowledgments

I am indebted to J. Finkelstein, V. Poénaru, and W. Swiatecki for advice in preparing this paper. I am also of course indebted to my dear friend, Francis Low, for having so felicitously arrived at the glorious age of 60. This work was supported by the Director, Office of Energy Research Office of High Energy and Nuclear Physics, Division of High Energy Physics of the U.S. Department of Energy, under contract W-7405-ENG-48. Support was also provided by the Miller Institute of the University of California.

References

[1] H. Harari, *Phys. Rev. Lett.* 22: 562 (1969); J. Rosner, *Phys. Rev. Lett.* 22: 689 (1969).

[2] G. Veneziano, *Nucl. Phys. B* 74: 365 (1974); *Phys. Lett. B* 52: 220 (1974).

[3] G. F. Chew and C. Rosenzweig, *Phys. Rep. C* 41, no. 5 (1978).

[4] D. Iagolnitzer, *The S Matrix*, North Holland, Amsterdam (1978).

[5] S. Mandelstam, *Phys. Rev.* 184: 1625 (1969); *Phys. Rev. D* 1: 1745 (1970).

[6] H. P. Stapp, *Phys. Rev.* 125: 2139 (1962); preprint LBL-10744, Berkeley (1981), submitted to *Phys. Rev.*

[7] G. F. Chew and J. Finkelstein, *Z. Physik C* 13: 161 (1982)

[8] G. F. Chew, J. Finkelstein, R. E. McMurray, Jr. and V. Poénaru, *Phys. Lett. B* 100: 53 (1981); *Phys. Rev. D* (October 15, 1981).

[9] G. F. Chew and V. Poénaru, *Phys. Rev. Lett.* 45: 229 (1980); *Z. Physik C* 11: 59 (1981)

[10] G. Weissmann, *Int. J. Th. Phys.* 17, no. 11: 853 (1978).

[11] P. Gauron, B. Nicolescu and S. Ouvry, *Phys. Rev. D* 24: 2501 (1981).

[12] G. F. Chew, *Phys. Rev. Lett.* 47: 754 (1981).

[13] L. Balazs, J. Finkelstein, and R. Espinosa, private communication, Berkeley (1981).

[14] G. F. Chew, J. Finkelstein and M. Levinson, *Phys. Rev. Lett.* 47: 767 (1981).

[15] G. F. Chew, J. Finkelstein, and V. Poénaru, *Phys. Rev. D* 24: 2764 (1981).

The Fixed Point of Classical Dynamical Evolution and Chaos

Mitchell J. Feigenbaum

Dynamical systems—for example fluids obeying the causal Navier-Stokes equations—undergo a transition to chaotic behavior that can be quantitatively understood through a nonperturbative method in the context of few degrees of freedom. An introductory discussion explains this connection and is followed by a presentation of the renormalization-group theory of period doubling.

Introduction

It is never very difficult to find dynamical systems whose behavior is sufficiently complex as to be intractable to all available methods of analysis. For example, a driven damped oscillator—Duffing's oscillator—is just such a system:

$$\ddot{x} + k\dot{x} + x^3 = b\cos 2\pi t. \tag{1}$$

(The potential in (1) is taken as pure quartic: the harmonic part does not modify the analysis, while to be an "impossible" problem, the non-linearity is crucial.)

To understand what (1) can do, it is important first to relate the problem specified by a differential system to that of a (discrete) map. First of all it is useful to identify the dynamical degrees of freedom of (1). There are many objects which enter the system through a first time derivative, so that to have a unique solution, the initial value of these quantities must be specified. Equation (1) has two such degrees of freedom:

$$\dot{x} = p,$$
$$\dot{p} = -kp - x + a\cos 2\pi t. \tag{2}$$

Such a system is termed nonautonomous, since it possesses explicit time dependence. Indeed, had (2) been autonomous, although its solution would be difficult to precisely determine, that asymptotic (in time) solution would be either static or periodic. (This is the content of the Poincaré-Bendixson theorem, and is true for all autonomous systems of two degrees of freedom.) A third autonomous degree of freedom or time dependence is the necessary ingredient to make (2) "impossible." (We could write (2) as an autonomous system of three degrees simply by defining

$$\dot{z} = 2\pi, \quad z(0) = 0.)$$

A solution to (2) is seen to be a trajectory in the x, p plane. Because of the time dependence, this trajectory *can* be self-intersecting: initial conditions for (2) require specifying x, p *and* the initial phase. Now what do these trajectories look like? Evidently, for sufficiently weak driving, the system sympathetically oscillates at the frequency of the drive, and its trajectory is simply a closed loop about the origin. A simple way to present this fact is to sample the trajectory at the clock rate specified by the drive. Then all that appears is a unique point in the x, p plane: that initial value (since the sampling fixes the phase) that the system periodically revisits each cycle. Had we started with initial conditions different from this periodic value, we would have seen a sequence of points converging to this limiting one. Accordingly, following the entire trajectory is a wasteful effort. But the dynamics *has* been specified differentially. What is desired is the implied *discrete* dynamics that determines that unique point which follows from a specified initial point after one clock cycle. This rule is then a mapping of the x, p space into itself, and for obvious reasons, is called the "time one map" of the system. To produce it, all we need do is specify x, p at a definite phase of the drive, and then integrate the system (2) for one second: the resulting values of x and p are exactly the image of the initial point under the time one map.

In this mapping language, the limiting point of the periodic trajectory is precisely a *fixed point* of the map. If we knew the map analytically, we could algebraically compute this value: in any case it is the procedure one must follow to find the limit cycle.

Imagine now that as we increase a that the system executes two loops—that is, requires two seconds to be periodic. Then the map would have a *pair* of limit points that would be visited alternately. To compute such a cycle, we must then find a fixed point of the time two map—which is just the time one map composed with itself. It thus follows that *any* periodic solution to (2) is obtained as a fixed point of an appropriate iterate of this map.

Should the system possess an *aperiodic* solution, the limit set would be infinite, corresponding to a trajectory that would become space-filling. Indeed, (2) possesses such "chaotic" solutions, and it is these solutions which are of a complexity that surpasses the power of traditional methods. Should the set of points for the time one map fill out a "solid" part of the x, p plane, then usual statistical methods could possibly be employed at least to compute various moments, correlation functions, etc. However this is basically *not* the case for (2). The reason that this is so is that (2) is a *dissipative* system. In fact, the flow (2) contracts phase space at the uniform rate

$$\frac{\partial \dot{x}}{\partial x} + \frac{\partial \dot{p}}{\partial p} = -k, \tag{3}$$

so that the time one map has a uniform Jacobian determinant

$$b = e^{-k}. \tag{4}$$

Since (2) can be reversed in time, each point has a unique pre-image under the map. Also, the solution of a differential equation is a smooth function of initial conditions. Thus, the time one map of (2) is a differentiable map with differentiable inverse, i.e., a diffeomorphism on the plane with a constant contractive Jacobian determinant. Indeed, the quadratic map on the plane

$$x' = y - x^2,$$
$$y' = a - bx, \tag{5}$$

known as Hénon's map, is just such a diffeomorphism and, not surprisingly, behaves altogether similarly to the time one map of (2). Since we *can't* integrate (2) exactly, all we know is that the map is something like (5).

Now, since the map of (2) contracts areas at the uniform rate (4), it immediately follows that any asymptotic set of points must have zero measure in the plane. This might mean, for example, that a region of finite measure (called the *basin*) contracts down onto a *finite* set of points (i.e., a periodic *attractor* with the determined basin of attraction). Or it might mean that an open set contracts down to an infinitely sheeted curve with zero measure but of a *fractional* (scaling) dimension between one and two. Such a set is called a "strange" attractor: the action of the map on the attractor is to *stretch* distances, while a strong contraction acts transversely down onto the attractor. In such a (usual) circumstance, nearby points on the attractor exponentially separate, so that slight errors in the preparation of an initial state lead to radically divergent future evolutions. This phenomenon of "sensitive dependence on initial conditions" is a feature that is troublesome and characteristic of turbulence and chaos. Also, since the attractor is so sparse (nowadays called a "fractal"), a correct statistical treatment of the system requires a probability distribution (i.e., measure) concentrated on a peculiar set, so that an a priori (say Gaussian) measure on all of phase space will yield seriously erroneous results. This then is a conceivable explanation of the failure of usual techniques in describing turbulence. (We are, however, assuming that no external noise couples to our totally *causal* system. How noise affects these considerations is largely open—although in some circumstances it is understood, and when weak enough is largely irrelevant.)

Accordingly, we are now faced with the questions of just what these strange attractors look like, how dynamics on them proceeds, and how they come into existence borne out of the more elementary periodic behaviors that typically occur for weak stressings of physical systems. We are, at the moment, rather ignorant of the answers to the first two questions. However, we have by now acquired some serious information pertinent to the last—or onset—problem. In exact quantitative ways (2) and (5) possess *identical* onset behavior. Indeed this *universality* is quite strong as it also embraces the onset problem in physical fluid-dynamical systems —this is a system of an *infinite* number of degrees of freedom. This universal behavior is the theme of this paper.

As we have said, it is important to pay attention to the map which (discretely) advances the system a characteristic *finite* interval of time, rather than to the original flow which advanced time infinitesimally. For a periodically driven system, this characteristic interval is specified, and an N-dimensional flow determines an N-dimensional time one map. In the case of an autonomous system, one must still find a map for which a fixed point corresponds to a cycle. Now, should the system be executing an oscillatory motion, then any "typical" coordinate of the system will serve as a clock: in particular the coordinate will pass through a local maximum value once per (approximate) cycle, so that the system to be observed is measured each time $\dot{x}_N = 0$, where x_N is a typical coordinate. Since the equations of motion include

$$\dot{x}_N = f_N(x_1, \ldots, x_N), \tag{6}$$

all we need do is sample the system whenever its N-dimensional trajectory crosses the "surface of section" $f_N = 0$ in a given sense (so that $\dot{x}_N = 0$ is a maximum and not a minimum). In this way, then, we are sampling the system at characteristic (though variable!) time intervals, and obtain an $(N-1)$-dimensional "Poincaré map." That is, $x_1, \ldots, x_{N-1}$ are specified;

x_N determined so that $f_N = 0$; the system of equations is integrated until f_N again vanishes with the trajectory piercing the surface of section in the same sense as it initially departed; and the new values $x_1, \ldots, x_{N-1}$ are returned as the image of the initial set under the Poincaré map.

Accordingly, much of the study of high-dimensional nonlinear flows is already contained in the study of high-dimensional and smooth nonlinear maps. Moreover, if the flow is *dissipative*, then the asymptotic dynamics lives in a lower-dimensional subspace of the original phase space. In fact, since the contractions (or expansions) are generically different in each direction, the local behavior of the long-term evolution of the system can be, and often is, one-dimensional. (Of course, since a contraction has led down to this 1D behavior, the resulting 1D map is now no longer invertible.) Thus we arrive at the fact that for certain ranges of an external parameter, a large system has its long-term behaviors determined by mappings of low dimensionality. This low-dimensional regime corresponds to the onset regime—when a system is making a transition from simple coherent motion to motions exhibiting the independence of its many modes. In particular, onset behaviors corresponding to 1D maps can be expected to have a restricted variety of possibilities, and accordingly we study such maps to see what such fluids are permitted to do.

1D Maps on the Interval

The question addressed is: What is the long-term behavior of maps on an interval? More precisely, we envisage 1-parameter families of such maps and ask how the asymptotic behavior depends upon the parameter. What we have in mind is that for small parameter values, the system relaxes down to a static solution; grows periodic as the parameter increases; until at some critical value, a transition to chaotic behavior occurs. That is, we are taking for granted (the fact) that even 1D discrete dynamics possesses chaotic solutions, and want to know what properties characterize the transition of such systems. Any scenario that emerges, then, also applies to high-dimensional dissipative flows. So, how do things work out?

First, some terminology. One step of evolution corresponds to applying the map once, while n steps requires n iterations of the map:

$$x_n = f^n(x_0); \quad f^{n+1}(x) = f^n(f(x)); \quad f^0(x) = x. \tag{7}$$

A static or periodic (of period 1) solution is

$$x_{n+1} = x_n; \quad \text{or} \quad x_n = x^* \quad \text{where} \quad x^* = f(x^*). \tag{8}$$

That is, the simplest periodic behavior is determined as a fixed point of the map f. A more complicated periodic solution of period p is a set of p points each satisfying

$$x^*_{n+1} = f(x^*_n); \quad x^*_{n+p} = x^*_n, \ n = 0, 1, \ldots, p-1; \quad \text{or} \quad x^*_n = f^p(x^*_n), \quad (9)$$

i.e., each is a fixed point of f^p.

Now, while (8) or more generally (9) determines for a given f definite values of the elements of certain cycles (i.e., for those p's and x^*'s for which real solutions exist), there is no reason to believe that such cycles represent the asymptotic behavior. For this to be true, it must be that the cycle is stable. That is, should x be near an x^*_n, then the iterates of x must converge towards x^*_n. Calling f^p F, we are requiring that with

$$x^* + \xi_{n+1} = F(x^* + \xi_n), \tag{10}$$

then $\xi_n \to 0$. With F smooth, we have in linear approximation

$$x^* + \xi_{n+1} \approx F(x^*) + \xi_n F'(x^*) = x^* + \xi_n F'(x^*),$$

and hence

$$\xi_n \approx \xi_0 [F'(x^*)]^n. \tag{11}$$

Thus if $|F'(x^*)| < 1$, then the iterates geometrically converge to the fixed point at the rate $\mu = F'(x^*)$. For periodic cases with $p > 1$, an elementary consequence of the chain rule is that

$$F'(x^*_n) = \frac{d}{dx} f^p(x^*_n) = \prod_{m=0}^{p-1} f'(x^*_m) \quad \text{for every} \quad n = 0, 1, \ldots, p-1, \tag{12}$$

so that each element of the orbit possesses the identical *stability* μ. Accordingly, if (9) possesses a solution, (12) provides a definite procedure to establish whether or not the solution corresponds to the asymptotic behavior of the system. In particular, a p-cycle with any stability μ for $|\mu| < 1$ is an asymptotic solution. This now leaves the question: For which p does (9) possess a solution, and which of these solutions are stable?

It is first of all clear that a *monotone f* can only have a set of fixed points (i.e., period 1) which are alternatingly stable or unstable. (A Poincaré map of an autonomous system of two degrees of freedom is a 1D *invertible* map, and so monotone. Thus, any such system has asymptotic solutions no more complicated than static or periodic motion of period 1—i.e., a limit cycle. This is why it is necessary to force such a system to get interesting behavior.) To get higher periods and chaos, the map must be *non*invertible, and hence fold an interval back onto itself. It is this folding, with the attendant possibility of stretching, which creates chaotic solutions. And by what has just been said, such maps can only arise with at least three degrees of freedom and dissipation. To proceed further, it is useful to have a model map in mind: we take the "logistic map"

$$x_{n+1} = 4\lambda x_n(1 - x_n), \tag{13}$$

which for $0 \le \lambda \le 1$ maps $[0, 1]$ into itself. Observe, at this point, that a parameter λ, has appeared: the solutions to (9) depend upon λ, and it is precisely the problem of how solutions to (9) depend upon λ that interests us.

To begin to study this question, observe first that for $0 < \lambda < \frac{1}{4}$ the fixed point at $x = 0$ is stable. At $\lambda = \frac{1}{4}$, $\mu = 1$. For $\lambda > \frac{1}{4}$, $x = 0$ is unstable, while a new fixed point in $[0, 1]$ has become stable. This situation persists until $\lambda = \frac{3}{4}$, when the stability has evolved from $+1$, at $\lambda = \frac{1}{4}$ when the fixed point arose, to -1. For $\lambda > \frac{3}{4}$ no fixed point is stable. Thus, for $\frac{1}{4} < \lambda < \frac{3}{4}$, the asymptotic behavior is known: the system relaxes to a static behavior (or a limit cycle of period 1), marginally stable at $\lambda = \frac{1}{4}$, growing more stable, until $\mu = 0$ (*superstable*) at $\lambda = \frac{1}{2}$, and then growing less stable, becoming marginal again at $\lambda = \frac{3}{4}$. When a simple limit cycle is unstable ($\lambda > \frac{3}{4}$), what then is the asymptotic solution to the system?

To state the answer as a fact, for a range of $\lambda > \frac{3}{4}$, a 2-cycle bifurcates into existence, and is stable. (At $\lambda = \frac{3}{4}$, the equation $f^2(x) = x$ has a degenerate third-order root: a complex conjugate pair of roots has collided with the marginally stable fixed point. For $\lambda > \frac{3}{4}$, the fixed point is unstable, and the previously complex pair are now the elements of a stable 2-cycle.) By the chain rule, if

$$f(x^*) = x^* \quad \text{then} \quad Df^2(x^*) = [f'(x^*)]^2.$$

Since $f'(x^*) = -1$ at $\lambda = \frac{3}{4}$, the newly born 2-cycle has at its emergence $\mu = (-1)^2 = +1$. As λ is further increased, one of the elements of the cycle moves away from the (unstable) fixed point toward $x = \frac{1}{2}$, where $f' = 0$. At some λ, $x = \frac{1}{2}$ becomes an element of the 2-cycle, and $\mu = Df^2(x^*) = 0$ by (12), so that a superstable 2-cycle exists. As λ is further increased, this element of the cycle moves into $(0, \frac{1}{2})$, and $\mu < 0$, until at some value of λ, say Λ_2, $\mu = -1$ and the 2-cycle in turn has become marginally stable.

At this point the system no longer possesses an asymptotic 2-cycle, and must become more complex. In terms of f^2, we have encountered a range of λ for which μ has varied from $+1$, became superstable, and then suffered loss of stability at $\mu = -1$, just as had been the case for f for a previous range of λ. Indeed, $f^4(x) - x$ has a triple zero at Λ_2, the emerging new zeroes being elements of a 2-cycle for f^2 or a 4-cycle for f, and, by the chain rule,

$$\mu = Df^4(x^*) = [Df^2(x^*)]^2 = +1.$$

Again there is a range of λ with $|\mu| < 1$, $\mu = 0$ at some value of λ, and finally at Λ_3, $\mu = -1$ and for $\lambda > \Lambda_3$ the 4-cycle is unstable. Inductively, this period doubling recurs ad infinitum, with a 2^n-cycle stable for $\Lambda_n < \lambda < \Lambda_{n+1}$; $\mu(\Lambda_n) = +1$, $\mu(\Lambda_{n+1}) = -1$, and $\mu(\lambda_n) = 0$. These Λ_n's are a monotone sequence strictly bounded by $\lambda = 1$, and so accumulate at a definite value $\lambda_\infty < 1$. At λ_∞ *no* periodic orbit is stable; rather, the system asymptotically moves aperiodically on a Cantor set, and we have an elementary prototypic "strange attractor." (As a limit of periodic attractors, motion on this set does not separate exponentially: it is a transitional behavior that just destroys periodicity and ushers in a regime where sensitive dependence on initial conditions can occur.) The scenario of this route, then, is periodic behavior of increasing (doubled) periodicity, until at λ_∞ the system has made a transition through the infrared to chaos: at λ_∞ the Fourier spectrum has marginally become continuous.

Now it turns out that not just the logistic map, but *any* 1D map with a unique extremum in an interval displays this identical period-doubling behavior as the "height" of the extremum is increased. (It is actually somewhat more general: any smooth λ-dependent change of coordinates preserves this phenomenon.) Thus, from the general considerations of the introduction, we can expect systems of high numbers of degrees of freedom (e.g., real fluids) to also undergo such a scenario as an appropriate parameter is varied. (As we shall see, this is in fact true.)

So far, these facts are of a qualitative nature. It turns out, though, that rigorous renormalization-group arguments establish universal quantitative features of this scenario that can readily be put to experimental test. That is, we can make precise numerical predictions of the rate of onset of aperiodic behavior and the precise nature of the transition dynamics (i.e., predict numerically the observed power spectrum) of *any* system undergoing this scenario. To understand this, let us quote some numerical results for (13).

The qualitative mathematical fact is that for each n, (13) possesses a superstable 2^n-cycle at λ_n. It is easiest to focus on the superstable cycles, because by (12) a superstable 2^n-cycle is one which has $x = \frac{1}{2}$ as an element. Thus, the determination of λ_n is through the condition

$$F_n(\lambda) \equiv f_\lambda^{2^n}(\tfrac{1}{2}) - \tfrac{1}{2} = 0 \ (\lambda = \lambda_n). \tag{14}$$

(Observe that $\lambda_0, \lambda_1, \ldots, \lambda_{n-1}$ also satisfy (14): as $n \to \infty$, F_n is developing an essential zero at λ_∞. To determine λ_n, first obtain $\lambda_0 (= \frac{1}{2})$ as the unique zero of F_0. Then find the smallest zero of F_1 distinct from λ_0, which is λ_1. Proceed inductively.) Since f is quadratic, f^n is a polynomial of degree 2^n. Thus for example, to obtain λ_{10}, we need the zero of a polynomial of degree

$$2^{2^{10}} \approx 2^{1000} \approx 10^{300}. \tag{15}$$

This precisely underscores the nature of the problem at hand: while each time step is trivial to obtain (just quadratic), the long-term evolution is totally unwieldy in terms of this basic unit, and new ideas are required to attain the understanding of this dynamical regime. Very few such tools are extant. However, we are lucky here and the ideas of scaling and the renormalization group are precisely the requisite tools.

By (15), we cannot explore analytically, and so resort to numerical computation. The result is excellent: we discover that λ_n converges to λ_∞ in an asymptotically elementary fashion:

$$\lambda_\infty - \lambda_n \sim \delta^{-n} \tag{16}$$

for large n, with $\delta > 1$. In fact, for (13), $\delta \approx 4\frac{2}{3}$, so that cycles accumulate very quickly. Geometric convergence suggests a scaling that proves correct. In fact, should we perform the same numerical experiment for the map

$$x_{n+1} = \lambda \sin \pi x_n,$$

the geometric convergence of (16) again emerges. However, the great discovery is that

$$\delta = 4.669\,201\,6 \ldots \tag{17}$$

is *identical* for the two maps! Indeed, as a is varied in (5), a_n's are obtained for successive period doublings with δ again given by (17). And of course, the same is true for Duffing's equation (1), as well as for almost every system of oscillatory nonlinear differential equations studied that display chaotic solutions. Indeed the value of δ given by (17) is observable in physical fluid-flow experiments. Thus, very restrictive consequences follow from the idea of dissipation leading to effective 1D maps!

Let us now see how specific those results are, and how we can come to calculate (the critical exponent) δ. (δ depends upon only some ingredient: every map with a *quadratic* extremum yields (17). The theory which follows is true *whatever* the nature of the extremum, although the solution determines different values for each kind of extremum. (17) is, however, the only real case of interest, since any smooth map with an extremum has a quadratic extremum unless some nongeneric symmetry is in force.) It is necessary to do one further piece of numerical exploration to discover *what* scaling leads to (16).

Recall that with each period doubling at Λ_n, the previously stable fixed points throw off a pair of almost coincident elements of the doubled cycle. Thus, just past the bifurcation point Λ_n, 2^n iterates almost return one element to itself, but actually map it into the other of the pair. Since λ_n is converging, these splitting distances must decrease with n. So at λ_{n+1}, 2^n iterations of the critical point (where $f'(x) = 0$) map it onto the now finitely separated other element of the pair thrown off at Λ_n. Numerically, it turns out that

$$f_{\lambda_{n+1}}^{2^n}\left(\tfrac{1}{2}\right) - \tfrac{1}{2} \sim (-\alpha)^{-n} \tag{18}$$

where α is another universal constant:

$$\alpha = 2.502\,907\,875\,\ldots. \tag{19}$$

For the sequel, it is convenient to perform a coordinate transformation that sends the critical point to $x = 0$, so that (18) becomes

$$\lim_{n\to\infty} (-\alpha)^n f_{\lambda_{n+1}}^{2^n}(0) = \text{finite}. \tag{20}$$

However, recall that for $\Lambda_n < \lambda < \Lambda_{n+1}$, $Df^{2^n} = \mu$ and $|\mu| < 1$. To obtain this finite derivative, it is clear that we should consider

$$\lim_{n\to\infty} (-\alpha)^n f_{\lambda_{n+1}}^{2^n}(x/(-\alpha)^n)$$

which, then, by (20), has a limit at $x = 0$, while for each n its derivative is also finite at $x = 0$. By a definite magnification m, we can adjust $f_m(x)$:

$$f_m(x) \equiv mf(x/m),$$

so that the limit in (20) is unity. Thus we find numerically that

$$\lim_{n\to\infty} (-\alpha)^n f_{\lambda_{n+1}}^{2^n}(x/(-\alpha)^n) = g_1(x) \tag{21}$$

exists, and is also *universal*. But (21) asserts that we can write down a priori the long-term behavior of any system following this scenario! That is, the transition dynamics are known.

To understand (21) observe that by our initial magnification by m, $g_1(0) = 1$ while also $g_1(1) = 0$ by the definition of λ_{n+1}. Thus $g_1(x)$ is a universal function with extremum at $x = 0$ possessing a 2-cycle. Each element of the 2^{n+1}-cycle at λ_{n+1} becomes an element of a 2-cycle for f^{2^n}. Since g_1 is universal, this means that the elements of the cycle are all universally determined through all the elementary 2-cycles specified by g_1, but in terms of the one-scale set by normalizing f. Thus the dynamics are truly universally determined. (This is an overstatement: the convergence to g_1 is such that at the nth approximation in (21), 2^r elements near $x = 0$ are located as exactly as desired for $r \ll n$. As n diverges, infinitely many points are exactly located, but are arbitrarily close to $x = 0$: this is an asymptotic theory, and arbitrarily accurate results follow from starting data for sufficiently large n. How this is employed to compute *all* of the dynamics is beyond the scope of this paper, and has been published elsewhere [1].)

So we are now left with the question of computing not only δ, but also the universal function g_1 and these computations are to be performed in a way independent of the iteration of any particular 1-parameter family of maps.

The Period-Doubling Renormalization Group

At this point, the discussion will grow more technical. At λ_{n+1},

$$f_{\lambda_{n+1}}^{2^n}(0) \sim (-\alpha)^{-n}.$$

Since $\lambda_n \approx \lambda_{n+1}$, then

$$f_{\lambda_{n+1}}^{2^{n-1}}(0) \sim (-\alpha)^{-n+1},$$

etc. (*Higher* iterates are required to discriminate between λ_n's for higher values of n.) Thus it can be guessed that at λ_∞, a scaling at all orders exists:

$$f_{\lambda_\infty}^{2^n}(0) \sim (-\alpha)^{-n}, \quad n \sim \infty. \tag{22}$$

With a suitable scaling of f, we then surmise that

$$\lim_{n \to \infty} (-\alpha)^n f_{\lambda_\infty}^{2^n}(x/(-\alpha)^n) = g(x), \quad g(0) \equiv 1, \tag{23}$$

exists with $g(x)$ universal. Observe that (23) singles out λ_∞ as a special isolated parameter value: for $\lambda = \lambda_N \; N \gg 1$, the $N - 1$ approximation in (23) would in fact be an approximation to g_1 which possesses a 2-cycle as opposed to the aperiodic behavior of g implied by (22). That is, for λ arbitrarily close to λ_∞, successive approximations to the limit of (23) at this fixed value of λ will ultimately diverge away from g. The divergence is very strong, and indeed, as will shortly be demonstrated, is exactly δ. Since, however, at λ_∞ the limit does exist, we are in a position to determine the equation it obeys:

$$(-\alpha)^{n+1} f_{\lambda_\infty}^{2^{n+1}} (x/(-\alpha)^{n+1})$$

$$= (-\alpha)(-\alpha)^n f^{2^n} \left(\frac{1}{(-\alpha)^n} (-\alpha)^n f^{2^n} \left(\frac{1}{(-\alpha)^n} x/(-\alpha) \right) \right).$$

Taking the limit as $n \to \infty$, by (23), we find

$$g(x) = -\alpha g(g(-x/\alpha)) \equiv T[g](x). \tag{24}$$

That is, $g(x)$ is the fixed point of an operator T, in *function* space: we can solve our problem only in the context of an *infinite*-dimensional space. In terms of T, the limit of (23) is simply

$$\lim_{n \to \infty} T^n[f_{\lambda_\infty}] = g. \tag{25}$$

By the above argument for $\lambda \approx \lambda_\infty$, we see that there are functions arbitrarily near to f_{λ_∞} such that the limit in (25) does not exist—or at least is not g. That is, g is an unstable fixed point of the operator T.

Before proceeding with the study of the instability of g, let us consider the solution to the fixed-point equation (24). Since g is supposed to be universal, we could obtain it through (25) by iterating

$$f_{\lambda_\infty} = a_\infty - x^2, \tag{26}$$

which is the logistic map after the change of coordinates that sends $x = \frac{1}{2} \to x = 0$. Since (26) is a function of x^2, so too are its iterates, so that the universal limit g is a symmetric function. With the specification that it is a function of x^2, the solution to (24) will yield α and g appropriate to the generic quadratic extremum. A solution to (24) which is a function of $|x|^z$ similarly determines α and g for the other, nongeneric universality classes. To see that α is determined, using the normalization $g(0) = 1$, and setting $x = 0$ in (24), we have

$$1 = g(0) = -\alpha g(g(0)) = -\alpha g(1) \to (-\alpha)^{-1} = g(1),$$

so that (24) is an equation solely for the function $g(x)$. Specifying that the solution is an analytic function of $|x|^z$ then determines a unique solution. (This is almost right. Correctly, there is an isolated solution to (24) of this form with precisely one eigenvalue of DT outside the unit circle.) Comparing coefficients of powers of $|x|^z$ up to a certain order and truncating higher-order terms is one method of producing an approximate solution, which can be done by hand for the lowest orders of approximation. Better methods are published elsewhere. It is important that (24) requires a perturbation expansion to offer a solution: this is a reflection on the state of available mathematical machinery. After all, the *form* of (24) is the most compact and precise encoding of a scaling structure: literally it asserts that the behavior for $2N$ steps is precisely a rescaling of the behavior for N steps, so that the solution to (24) is self-similar to all scales.

We now face the question of g's stability. This is the same sort of problem presented by (10), with the significant difference that the context has now grown many more dimensions. Comparing (21) and (23), what we want to know is the asymptotic behavior of

$$T^n[f_\lambda](x) = (-\alpha)^n f_\lambda^{2^n}(x/(-\alpha)^n) \tag{27}$$

which we can now analyze near T's fixed point. Since, by (23),

$$T^n[f_{\lambda_\infty}] \sim g$$
$$T^n[f_\lambda] = T^n[f_{\lambda_\infty} + (\lambda - \lambda_\infty)\partial_\lambda f + \cdots] = T^n[f_{\lambda_\infty}] + (\lambda - \lambda_\infty)DT^n \cdot \partial_\lambda f$$
$$+ \cdots \sim g + (\lambda - \lambda_\infty)(DT_g)^n \cdot \partial_\lambda f + \cdots . \tag{28}$$

In (28), the linear operator DT^n is the functional derivative of T^n at f_λ, which is

$$DT\big|_{T^{n-1}f_{\lambda(\infty)}} \cdot DT\big|_{T^{n-2}f_{\lambda(\infty)}} \cdots DT\big|_{f_{\lambda(\infty)}} \sim DT_g \cdots DT_g = [DT_g]^n \tag{29}$$

where DT_g is the derivative of T at its fixed point g. What is more precisely meant by (28) and (29) is that for $n \sim \infty$, we are projecting out the eigenfunction of DT with largest eigenvalue. That is, if δ is the largest eigenvalue of DT_g with eigenfunction Ψ_δ, then

$$DT^n\big|_{f_{\lambda(\infty)}} \cdot \partial_\lambda f \sim \delta^n \psi_\delta(x)c(f), \quad \psi_\delta(0) \equiv 1, \tag{30}$$

where the dependence on f is through the projection constant $c(f)$. That is, returning to (28),

$$T^n[f_\lambda] \sim g + (\lambda - \lambda_\infty)\delta^n\psi_\delta c(f). \tag{31}$$

Equation (31) is rich in content. First, setting $\lambda = \lambda_n$, $x = 0$, we have

$$0 = T^n[f_{\lambda_n}](0) \sim 1 + (\lambda_n - \lambda_\infty)\delta^n c(f) + \cdots, \tag{32}$$

which implies that

$$\lambda_\infty - \lambda_n \sim k(f)\delta^{-n},$$

so that (16) is now justified, while we see that (17) is now computable by solving DT's eigenvalue problem. (In passing, the explicit form of DT_g is

$$DT_g[\psi](x) = \alpha[\psi(g(x/\alpha)) + g'(g(x/\alpha))\psi(-x/\alpha)]. \tag{33}$$

With g available to a certain approximation from solving (24) we can numerically obtain the spectrum to similar accuracy. The interested reader can discover that the nongeneric case of extremum $|x|^{1+\varepsilon}$ provides an ε-expansion setting for solving (24) and the spectrum of DT_g.)

With (16) established, next set $\lambda = \lambda_{n+r}$ in (31):

$$T^n[f_{\lambda_{n+r}}] = (-\alpha)^n f_{\lambda_{n+r}}^{2^n}(x/(-\alpha)^n) \sim g - k(f)c(f)\delta^{-r}\psi_\delta,$$

i.e.,

$$g_r \equiv \lim_{n\to\infty} (-\alpha)^n f_{\lambda_{n+r}}^{2^n}(x/(-\alpha)^n) \sim g - k(f)c(f)\delta^{-r}\psi_\delta, \tag{34}$$

so that a host of universal functions related to g_1 of (21) exist as asymptotic limits, and are determined for large r by (34). Like g_1, all the functions g_r locate cycle elements: by (34),

$$g_r^{2^r}(0) = 0, \tag{35}$$

so that g_r is a universal function possessing a superstable 2^r-cycle. That is, g_r organizes the elements of asymptotically large cycles at a magnification of 2^r "points per bump" of g_r. It also follows from (34) and (35) that $c(f)k(f)$ is a determined universal constant k, so that (34) finally reads

$$g_r \sim g - k\delta^{-r}\psi_\delta. \tag{36}$$

(In fact (36) suggests the 1-parameter family of maps

$$g_\lambda = g - \lambda\psi_\delta.$$

This family period doubles, and, by (35), $\lambda_r \approx k\delta^{-r}$.) Since we now know that the limits g_r exist, the same manipulations that led to (24) now produce

$$g_{r-1} = T[g_r], \tag{37}$$

so that once we have entered the sequence $\{g_r\}$ for large r by (36), we then apply T exactly r times to enter the nonlinear regime, and obtain g_0, which locates cycle elements as fixed points at g_0's extrema. Thus all the dynamics at $\lambda \approx \lambda_\infty$ are available in a universal fashion.

It should be clear at this point that what we have is renormalization-group treatment of normal critical-point behavior. The space of functions on which T acts is identified with the space of Hamiltonians. A given interaction—i.e., with specified coupling constants—yields a 1-parameter family of functions parametrized by the temperature. T is the renormalization-group operator. By (25), at the critical temperature λ_∞, the 1-parameter ray crosses the critical surface (mathematically, the stable manifold through g): the Hamiltonian is critical, and successive applications of T move it into the fixed-point Hamiltonian g. Since just one parameter is sufficient to adjust to criticality—in (31), setting $\lambda = \lambda_\infty$ kills the unstable mode: DT_g must have just one eigenvalue outside the unit circle, i.e., one relevant eigenvalue—the situation is that of a normal second-order transition. Thus the behavior of 1D maps on the interval has provided us with a rigorous renormalization group in closed form, and we have understood the reasons for universality.

In conclusion, we now see that we have a theory with *no free parameters* that should quantitatively describe a host of high-dimensional dissipative systems on their route to chaos. For example, by this point several convection experiments have demonstrated five levels of period doubling with δ, and some dynamical predictions (usually power spectra) have been measured and found to be in agreement with theory at the 10% level [2]. (The interested reader is referred elsewhere to learn how Fourier spectra are calculated from this theory [1].) Although we have treated only the periodic side of the approach to the fixed point, observe that (31) is applicable to *all* $\lambda \approx \lambda_\infty$, so that early chaotic behavior is also determined. Again, details have been published elsewhere [3]. Finally, the spirit of these ideas has been literally applied (1) to rigorously extend 1D behavior to N-dimensional dissipative contexts [3], (2) to compute the disappearance of islands in 2D conservative context [4], and (3) to understand the transition to chaos in quasiperiodic flow. (The last is under heavy investigation at the moment.) In the future we can hope to understand the fractional-dimensional strange attractors that appear after the system has issued out of the 1D onset regime.

References

[1] M. J. Feigenbaum, *Phys. Lett. A* 74: 375 (1979); *Comm. Math. Phys.* 77: 65 (1980); *Nonlinear Phenomena in Chemical Dynamics*, Springer-Verlag (1981), pp. 95–102.

[2] A. Libchaber and J. Maurer, *J. Phys. Colloq.* 41: C3–51 (1980); M. Giglio, S. Muzzati, and U. Perini, *Phys. Rev. Lett.* 47: 243 (1981); A. Libchaber, C. Laroche, and S. Fauve, to appear in *J. Phys. Lett.* (1982); P. S. Linsay, *Phys. Rev. Lett.* 47: 1349 (1981).

[3] Also, as a general reference and for references to the earlier literature see P. Collet and
J.-P. Eckmann, *Iterated Maps on the Interval as Dynamical Systems*, Birkhaüser, Boston
(1980).

[4] J. M. Greene, R. S. MacKay, F. Vivaldi, and M. J. Feigenbaum, *Physica D* 3: 486
(1981); P. Collet, J.-P. Eckmann, and H. Koch, *Physica D* 3: 457 (1981). For a review and
the most extensive list of references on this subject, see R. H. G. Helleman, in *Fundamental
Problems in Statistical Mechanics*, vol. 5, edited by E. G. D. Cohen, North Holland (1980),
pp. 165–233.

The Thermal Conductivity of Metals and Insulators

Victor F. Weisskopf

1. Introduction

I have often bothered Francis with primitive questions in physics. He always was ready to discuss them with me; sometimes he may even have enjoyed it. One day I came and asked him why the thermal conductivity of a metal is so much higher than that of an insulator. Everybody who has touched a piece of metal or a piece of rock on a cold day knows this well.

Take as an example the metal Na, and the crystal NaCl. At room temperature the ratio of the thermal conductivities is 19.5. Why is this? Here it comes:

The expression for thermal conductivity is

$$\kappa = \tfrac{1}{3}Clv, \tag{1}$$

where C is the heat capacity and l and v are mean free path and velocity of the carriers of the heat: electrons or phonons respectively.

The heat capacity of a metal is much lower than that of an insulator. In the former only the electrons within $k_B T$ from the top energy ε_F of the Fermi distribution are free to take up heat energy, whereas in the crystal, all lattice vibrations can do it, if the temperature $T > \theta_D$, where θ_D is the Debye temperature. The heat capacity C is the derivative of the thermal energy E_T per cm^3 with respect to T. In a crystal the $3n$ lattice vibrations (oscillators) each carry the energy $k_B T$ when $T > \theta_D$ and n is the number of atoms per cm^3. Thus $E_T = 3nk_B T$. In a metal (one electron per atom in sodium) only $\approx nk_B T/\varepsilon_F$ electrons can carry heat. We thus get:

Crystal: $\quad C = 3k_B n,$

Metal: $\quad C = \dfrac{\pi^2}{2}\dfrac{k_B T}{\varepsilon_F}k_B n.$

The factor $\pi^2/2$ is the result of a more detailed calculation. The thermal conductivities become

Crystal: $\kappa = k_B n v_{ph} l_{ph}$ (2)

Metal: $\kappa = \dfrac{\pi^2}{6} \dfrac{k_B T}{\varepsilon_F} n v_{el} l_{el}$ $(T > \theta_D)$,   (3)

where the subscripts ph and el refer to phonons or electrons. Therefore the ratio of conductivities is

$$\frac{\kappa_{Na}}{\kappa_{NaCl}} = \frac{\pi^2}{6} \cdot \frac{k_B T}{\varepsilon_F} \cdot \frac{n_{Na}}{n_{NaCl}} \cdot \frac{v_{el}}{v_{ph}} \cdot \frac{l_{el}}{l_{ph}}.$$ (4)

Here are five factors. The first three factors give about 1/160 for room temperature. Now v_{el} is the Fermi velocity in Na and v_{ph} is the sound velocity in NaCl. Taking the known data, we get $v_{el}/v_{ph} = 233$. Putting this into (4) and equating it with the observed ratio 19.5 of heat capacities, we get

$$\frac{l_{el}}{l_{ph}} = 13.6.$$ (5)

We see from this result that the difference in the heat conductivities arises mainly from the difference in the mean free paths. If the ratio (5) were more of the order unity, the conductivities would be about equal since the low ratio $k_B T/\varepsilon_F$ of heat-carrying electrons is just about canceled by the ratio of velocities.

The fact that the mean free paths differ by more than an order of magnitude seemed to me difficult to understand, and I asked Francis for an explanation.

2. The Electron and Phonon Velocities

Before we discuss that question it may be fun to see how one can understand the magnitude of the ratio v_{el}/v_{ph}. We consider a metal to be a cubic lattice filled with a degenerate gas of free electrons, one per atom. The lattice distance d is of the order of a few Bohr radii. Therefore we observe the approximate identity:

$$\varepsilon_A \equiv \frac{e^2}{d} \approx \frac{h^2}{md^2} \approx \varepsilon_F.$$ (6)

Here m is the mass of the electron.[1] Clearly $\varepsilon_A \approx \varepsilon_F$ since the wavelength

1. Here and in the following the letter h stands for $\hbar$, Planck's constant divided by 2π.

of the electrons on top of the Fermi distribution is of the order d. ε_A is an atomic energy of the order of a few electron volts. The momentum of the fastest electrons on top of the Fermi distribution is

$$p_F \approx \frac{h}{d},$$

and therefore their velocity is

$$v_{el} = \left(\frac{2\varepsilon_F}{m}\right)^{\frac{1}{2}} \approx \frac{h}{md}. \tag{7}$$

The velocity is of the order of the speed of electrons in atoms.

We are going to describe the mechanics of the lattice in the simplest possible terms: every ion is an independent oscillator with mass M and frequency ω_D (the Debye frequency). Let us determine this frequency. We expect the potential energy $\frac{1}{2}M\omega_D^2 x^2$ of a displacement x to be $\approx \varepsilon_A$ if $x \approx d$. Hence we get from (6)

$$\frac{1}{2}M\omega_D^2 d^2 \approx \varepsilon_A \approx \frac{h^2}{md^2},$$

and we obtain immediately

$$h\omega_D \approx \left(\frac{m}{M}\right)^{\frac{1}{2}} \varepsilon_A. \tag{8}$$

Now ω_D is the frequency of the sound wave of the shortest wavelength $\lambda_{min} \approx d$. Its velocity is

$$v_{ph} \approx \omega_D d \approx \left(\frac{\varepsilon_A}{M}\right)^{\frac{1}{2}}.$$

Comparison with (7) gives

$$\frac{v_{el}}{v_{ph}} \approx \sqrt{\frac{M}{m}}. \tag{9}$$

This value is about 230 for Na and NaCl, whereas the data give 233. Too good for such approximate calculations!

3. The Mean Free Path of Electrons

Now we go back to (5) and learn about mean free paths of electrons and phonons in lattices. First let us be sure that a regular perfect crystal lattice at zero temperature does not scatter either electrons or phonons.

It is the same effect that makes a regular lattice transparent for light, except when the wavelength is a simple fraction of the lattice distance. In that case we get Bragg reflection. The same is true for electron waves. The conditions for Bragg reflections represent forbidden bands for the electrons. When these conditions are not fulfilled, electrons move in a perfect lattice like free particles with a somewhat different energy-momentum relation that is expressed by an effective mass different from the actual one. This difference is small in Na; so we neglect it here.

The mean free path is infinite in a perfect crystal. The most important reason for its actual finite value is the normal motion of the atoms. Let us first determine the electron mean free path. We argue this way: if the Na ion were isolated, the electron cross section would be

$$\sigma_0 = ad^2, \tag{10}$$

where d is the lattice distance and a some constant ≈ 1. Assuming that the scattering is roughly isotropic, the isolated ion produces a scattered wave with a wave vector $\mathbf{k}$ into the solid-angle element $d\Omega$:

$$\left(\frac{\sigma_0}{4\pi}\right)^{\frac{1}{2}} d\Omega \exp{(\mathbf{rk} \cdot \mathbf{r})}, \tag{11}$$

when the ion is located at $\mathbf{r} = 0$. When the ion is displaced by δ the scattered wave would be (11), but with $\mathbf{r}$ replaced by $\mathbf{r} - \delta$. In a perfect lattice the scattered wave (11) of a given ion is canceled by the effects of all other ions. But if one ion is displaced, there will be a residual scattered wave

$$\left(\frac{\sigma_0}{4\pi}\right)^{\frac{1}{2}} d\Omega [e^{ik(\mathbf{r} - \delta)} - e^{ik\mathbf{r}}] \approx \left(\frac{\sigma_0}{4\pi}\right)^{\frac{1}{2}} d\Omega \frac{\mathbf{k} \cdot \delta}{i} e^{ik \cdot \mathbf{r}}$$

when $|\delta| \ll |k|^{-1}$. Hence the effective cross section of a displaced ion in the lattice is

$$\sigma_{\text{eff}} = \sigma_0 \cdot \frac{k^2 \delta^2}{3},$$

where $k^2 \delta^2/3$ is the average of $(\mathbf{k} \cdot \delta)^2$. The thermal displacement δ is determined as follows: it takes an energy ε_F (defined by (6)) in order to displace an ion by βd, where β is a constant $\lesssim 1$. Since the ion is an oscillator with energy $3kT$, and the square of the displacement is proportional to the energy, we obtain

$$\frac{\delta^2}{d^2} \approx \frac{3k_B T}{\varepsilon_F}. \tag{12}$$

The mean free path is

$$l_{el} \approx \frac{1}{\sigma_{eff} \cdot n} \approx \frac{d^3}{\sigma_{eff}} \approx \frac{\varepsilon_F}{k_B T} \frac{d^2}{\sigma_0} \frac{1}{k^2 d^2} d,$$

d^2/σ_0 and $k^2 d^2$ are of order unity; and we get

$$l_{el} \approx \frac{\varepsilon_F}{k_B T} \cdot d, \tag{13}$$

several hundred times larger than d at room temperature.

4. Interlude: Electric Conductivity

Before we go on to discuss the mean free path of phonons it may be fun to use (13) for the determination of the *electric* conductivity of metals. An electric field E exerts a force eE in the direction of the field. The electrons acquire a drift velocity

$$w = \frac{e}{m} \tau,$$

where $\tau \approx l_{el}/v_{el}$ is the time between two collisions. Thus we get a drift current density $j = ewn$, where $n = d^{-3}$ is the number of electrons per cm^3. The conductivity c is defined by $j = cE$, so that with the help of (13) we get

$$c \approx \frac{e^2 n}{m} \frac{l_{el}}{v_{el}} \approx \frac{e^2 n}{m} \frac{\varepsilon_F}{k_B T} \frac{d}{v_{el}}. \tag{14}$$

We obtain by using (7) and (6): $d/v_{el} \approx h/\varepsilon_F$, so that

$$c \approx \frac{\omega_p^2}{4\pi} \frac{h}{k_B T} \tag{15}$$

where the plasma frequency ω_p is defined as $\omega_p^2 = 4\pi e^2 n/m$. The relation (15) is reasonably well fulfilled. It says that the conductivity—a reciprocal time[2]—is the geometrical mean between the plasma frequency and the time τ^{-1} between two collisions.

By comparing the first equality in (14) with expression (3) for the metallic thermal conductivity, we obtain the famous Wiedemann-Franz Law for the ratio κ/c. The mean free path l_{el} drops out and one gets by using $\varepsilon_F = \frac{1}{2} m v_{el}^2$:

---

2. It is the inverse of the time it takes for a charge accumulation to spread over a volume e times the initial one.

$$\frac{\kappa}{c} = \frac{\pi^2}{3e^2} k_B T \quad \text{(metals)},$$

a well-tested relation.

5. The Mean Free Path of Heat Conduction

Now we come to the question that I asked Francis Low: Why should the mean free path of a phonon be so different from that of an electron? We discussed the issue and could not find a good answer. Then he told me: "Why don't you ask your old friend Rudolf Peierls; he certainly knows the answer." The younger generation of physicists is not interested in simple questions. They come with propagators, virtual particles, and solutions in 3.9 dimensions and assure us that the computer output would give the right result anyway.[3] I followed his advice and received a letter from Rudi after a decent time interval. It did contain the essential points and it directed me to find what I think is the right answer. He should not be held responsible for what I now will demonstrate. He also quoted one of those untranslatable idiomatic expressions of Pauli, who said, when he was faced with a similar problem: "Es wird schon so sein müssen, sonst wär's ja nicht so." A very inadequate translation, that does not express the wit and the irony, would be: "There must be reasons for it, or it wouldn't be so."

The difficulty with phonons comes from three sources: One is the fact that the thermal disturbances of a perfect lattice are the phonons themselves. The second point is the fact that there is no phonon scattering when the lattice motion is purely harmonic. Then the sound waves are independent of each other; under this condition the ion displacements δ do not cause any phonon scattering. Hence only the deviations from harmonicity are responsible for the scattering. The third point is this: a mere scattering of one phonon by another does not necessarily limit the heat transport, since the momentum in a given direction is simply transferred to the other phonon. If the total phonon momentum is conserved, the heat transport would not be reduced by scattering. The heat would be transported with sound velocity, and there would be no dissipation of heat.

The reason why dissipation takes place and heat is not transported with sound velocity comes from the fact that the momentum is *not* always

---

3. The statement contained in that sentence is mine, not Francis's.

conserved when sound waves are scattered by one another. The lattice is able to take over momentum differences in multiples of h/d. Because of the discreteness of the lattice, a wave with a wave number $k > h/d$ is indistinguishable from a wave with $k' = k - h/d$. When a wave with $k > h/d$ is produced, it must not be counted as one with a momentum hk but with $hk - h/d$. The difference in momentum is taken up by the whole lattice. This process was discovered by R. Peierls and he christened it by an equally untranslatable name: "Umklapp-Prozess." Since the highest phonon has a momentum of order h/d, the lattice takes over momentum only if the momentum transfer is larger than h/d, that is, when phonons with momenta of at least $h/(2d)$ collide.

These points would all indicate a longer mean free path for phonons; only the anharmonicities produce scattering, and only the scattering of the high-frequency phonons is effective; moreover an umklapp process must occur. This is why I turned to Francis and then to Peierls for help.

In the discussion of the electron mean free path, we started with the scattering by an isolated atom. Here we cannot do so, since sound waves exist only in the lattice. Also we must keep in mind that only deviations from harmonicity cause a scattering. The exact calculations are rather complicated and opaque. In order to get a semiquantitative result we argue as follows. First, we consider the potential energy $V(\delta)$ of the displacement δ of an atom from its rest position. We set $V(0) = 0$. Since the deviations from anharmonicity are essential, we write

$$V(\delta) = \left(\frac{a}{d^2}\delta^2 + \frac{b}{d^3}\delta^3 + \cdots\right)\varepsilon_A, \tag{16}$$

where ε_A is an "atomic" energy as defined by (6), d is the lattice distance, and a, b are numerical constants. Since $V(\delta) \approx \varepsilon_A$ for $S \approx d$, we conclude that $a \approx 1$. We also expect that $b \approx 1$ since the deviations from harmonicity ought to be of the order of the harmonic (quadratic) terms when $\delta \approx d$. Let us call V_a the anharmonic part of $V(\delta)$:

$$V_a(\delta) \approx \left(\frac{\delta}{d}\right)^3 \varepsilon_A + \cdots. \tag{17}$$

We now make the following assumptions. Whenever the energy V_a is about equal or larger than the phonon energy $h\omega$, the phonon will be strongly scattered by the cell in which it happens. (A cell is a region of order d^3.) This assumption is based upon the following consideration: We are interested in the scattering of high-frequency phonons, $\omega \approx \omega_D$, since only those are subject to the umklapp process. If the displacement

δ of a lattice atom is such that $V_a(\delta) \approx \hbar\omega_D$, the cell around that atom acts upon the phonon like a region with a strongly altered refraction coefficient. Thus we expect the scattering cross section of a given cell to be $\sigma \approx d^2$, when $\delta = \delta_0$ where δ_0 is defined by $V_a(\delta_0) \approx \hbar\omega_D$. We then make the not unreasonable assumption that σ_{eff} is proportional to δ^2 for δ smaller than δ_0. Considering only the cubed terms in (17), we then find

$$\delta_0 \approx \left(\frac{\hbar\omega_D}{\varepsilon_A}\right)^{\frac{1}{3}} d \approx \left(\frac{m}{M}\right)^{\frac{1}{6}} d,$$

where the last equal sign follows from (8). We approximately determine the value of δ at a temperature T by neglecting the cubic terms. Then (12) is valid with $\varepsilon_F \approx \varepsilon_A$ and we get

$$\sigma_{\text{eff}} \approx d^2 \cdot \frac{\delta^2}{\delta_0^2} \approx \frac{3k_B T}{\varepsilon_A}\left(\frac{M}{m}\right)^{\frac{1}{3}} d^2.$$

The mean free path becomes

$$l_{\text{ph}} = \frac{d^3}{\sigma_{\text{eff}}} \approx \frac{\varepsilon_A}{k_B T} \cdot d \cdot \frac{1}{3}\left(\frac{m}{M}\right)^{\frac{1}{3}}.$$

This is considerably smaller than the corresponding value (13) for electrons. We would get

$$\frac{l_{\text{el}}}{l_{\text{ph}}} \approx \frac{1}{3}\left(\frac{m}{M}\right)^{\frac{1}{3}} \approx 100, \tag{19}$$

which is much larger than the value (5), which we wanted to explain. We got what we were looking for but we overshot our aim by an order of magnitude! It can easily be repaired, however, since we must take into account that only those phonons contribute to the dissipation of heat that are able to perform umklapp processes. Say, the number of those smaller by a factor $\gamma < 1$. Furthermore the probability that a collision among those leads to an umklapp is $\xi < 1$. Thus the thermal conductivity of a crystal is not given by (2) but by

$$\kappa = (\gamma\xi)^{-1} k_B n v_{\text{ph}} \cdot l_{\text{ph}}. \tag{20}$$

Using (20) instead of (2), the observed value of the ratio (4) gives us

$$\frac{l_{\text{el}}}{l_{\text{ph}}} = \frac{13.6}{\xi \cdot \gamma}, \tag{21}$$

after inserting the data for the relevant magnitudes. It is not implausible that $\xi \cdot \gamma \approx 1/10$. Thus our result (19) is in reasonable agreement with the empirical ratio (21).

The observed smaller thermal conductivity of a crystal compared to a metal can be understood in a qualitative way. The sound waves are scattered more strongly than electron waves by equal atomic displacements. It is because the phonon energy is much lower and, therefore, the sound waves are more sensitive to changes in the atomic environment. Of course, such considerations do not explain the quantitative relations. Our aim was comprehension, not explanation.

It is of interest to discuss one glaring exception to the rule that nonconducting crystals have a smaller thermal conductivity than metals. It is diamond. Its thermal conductivity at room temperature is 4.6 times that of sodium and almost twice that of copper![4]

It is easily seen why diamond is such an excellent heat conductor. The four outer electrons of carbon are in the well-known hybrid state, in which they form four prongs in the directions of the corners of a tetrahedron. This arrangement allows the formation of a hexagonal lattice where the prongs of neighboring atoms overlap very well and produce strong bonds. This is the cause of the extreme hardness of diamond. The existence of very strong bonds has two consequences relevant to the thermal conductivity. Firstly, the number n of atoms per cm^3 is 7.9 times higher than in NaCl. Diamond is the material with the highest number of atoms per unit volume![5] This number enters linearly into the expression (2) of the heat conductivity κ. Secondly, diamond has an abnormally high Debye frequency ω_D because of the very strong restoring forces against displacements of the atoms in the lattice. It is 6.95 times larger than that of NaCl. Whereas the Debye temperatures $T_D = h\omega_D/k_B$ of ordinary crystals, such as NaCl, are of the order of room temperature or less, T_D for diamond is 2230°. Thus, most phonons present at room temperature have frequencies considerably less than half the Debye frequency. They cannot perform umklapp processes, and therefore heat travels rather unhindered through diamond.

---

4. Roy Schwitters told me of an unexpected experience, caused by this property of diamond. He had to mount diamond crystals on a base of wax; when he pressed them into the heated wax he burnt his fingers! The high heat conductivity also makes diamond such an effective cutting tool since it dissipates the heat produced in the cutting process.
5. Ed Purcell drew my attention to this fact and to its relation to the abnormal heat conductivity of diamond.

A Simple Quantum-Mechanical Problem

Herman Feshbach

Introduction

The problem to be discussed in this paper was suggested by considerations of the phenomena which occur when a proton of relativistic energy strikes a heavy nucleus [1]. In one type of experiment, the angular distribution in the laboratory frame and average energy of the nuclear fragments produced by such a collision are observed. It is found that as the proton energy increases the angular distribution becomes more peaked in the forward direction. However, in the neighborhood of a few-GeV proton energy this pattern changes. As the proton energy increases the angular distribution becomes less forward peaked, until at 300 GeV it is isotropic. At 28 GeV, for example, the angular distribution is rather flat with a maximum/minimum ratio of about 2 with the maximum at 70°. Similarly it is found that the average fragment energy increases until the proton has an energy of a few GeV. Beyond this value, increasing the proton energy is accompanied by a decrease in the average fragment energy, reaching an apparently asymptotic value well before the proton energy of 300 GeV. In another set of experiments, the multiplicity of charged relativistic particles ($v/c \geq 0.8$) is measured for a variety of target nuclei. It is found for a variety of incident particles (π, p, K) that the multiplicity in the forward cone changes only by a factor of approximately two when the target nucleus is changed from hydrogen to lead; that is, cascading does not occur.

These phenomena have the immediate interpretation that much of the energy of the energetic incident proton is not being deposited in the nuclear degrees of freedom. Rather, the internal degrees of freedom of the incident nucleon are excited, the deexcitation taking place after the incident nucleon leaves the nucleus. One can picture the process taking place as follows. The incident nucleon upon colliding with a target nucleon will be excited. In free space this would be followed by a decay involving the emission of several pions, etc. Within the nucleus, however, this process is interrupted by a second collision with another nucleon of the target nucleus. If this occurs within a sufficiently short time, the decay will be inhibited and will not occur until the excited nucleon leaves the nucleus. The critical parameters are the mean free path λ of the nucleons inside the nucleus, about 2 fm at these energies, and the lifetime τ, of the excited state of the hadron. The latter is roughly 1 fm/c in the rest frame of the decaying system. This must be boosted because of time dilation. If we assume, and this is only rough, that the rest frame is identical with the center-of-mass frame of the colliding nucleons, that is, that all the kinetic energy goes into internal energy, one finds that

$$\tau_{\text{lab}} = \sqrt{\frac{E_{\text{lab}}}{2mc^2}}\,\tau.$$

The critical energy is given by the condition that

$$c\tau_{\text{lab}} \approx \lambda.$$

Inserting the values of τ and λ yields a critical energy of $8mc^2 \approx 7.5$ GeV, of the correct order of magnitude.

A Simple Problem

Here we examine the assertion that the decay of the excited nucleon is inhibited by collisions as postulated in the foregoing discussion. Toward this end we consider the following model. The incident nucleon is assumed to travel in a straight line through the nucleus with the light velocity c. It suffers collisions at times $t_1, t_2, \ldots, t_n$, where n is a small number of the order of the radius of the nucleus divided by the mean free path λ. Secondly we assume that the nucleon has only one excited state ψ_1, in addition to the ground state ψ_0. The excited state has a complex energy $E = \varepsilon - i\Gamma/2$. The ground state energy is taken to be zero. Finally it is assumed that the collisions are impulsive so that the Schroedinger equation becomes

$$i\partial\psi_0/\partial t = \mu \sum_i \delta(t - t_i)\psi_1(t^+), \tag{1}$$

$$i\partial\psi_1/\partial t = E\psi_1 + \mu \sum_i \delta(t - t_i)\psi_0(t^+), \tag{2}$$

where μ is a coupling constant. The initial conditions are $\psi_0(0) = 1$ and $\psi_1(0) = 0$. The nature of singularity at t_i is defined by the integrated form of these equations. For example, $i(\psi_0(t) - \psi_0(0)) = \mu \sum_i u(t - t_i) \psi_1(t_i^+)$ where $t_i^+ = t_i + 0^+$ and u is the unit function. Letting

$$\psi_1^{(n)} \equiv \psi_1(t_n^+),$$
$$\psi_0^{(n)} \equiv \psi_0(t_n^+),$$

one obtains the recurrence relations

$$\psi_1^{(n)} = \frac{1}{1 + \mu^2} [e^{-iE\lambda(n,n-1)} \psi_1^{(n-1)} - i\mu\psi_0^{(n-1)}], \tag{3}$$

$$\psi_0^{(n)} = \frac{1}{1 + \mu^2} [\psi_0^{(n-1)} - i\mu e^{-iE\lambda(n,n-1)} \psi_1^{(n-1)}], \tag{4}$$

where

$$\lambda(n, n - 1) = t_n - t_{n-1}. \tag{5}$$

Define the vector

$$\Psi^{(n)} = (1 + \mu^2)^{n/2} \begin{pmatrix} \psi_0^{(n)} \\ \psi_1^{(n)} \end{pmatrix}. \tag{6}$$

These equations may be rewritten for convenience as follows:

$$\Psi^{(n)} = M_{n,n-1} \Psi^{(n-1)}, \tag{7}$$

$$M_{n,n-1} \equiv e^{-i\phi/2} \begin{pmatrix} \cos\dfrac{\theta}{2} e^{i\phi/2} & -i\sin\dfrac{\theta}{2} e^{-i\phi/2} \\ -i\sin\dfrac{\theta}{2} e^{i\phi/2} & \cos\dfrac{\theta}{2} e^{-i\phi/2} \end{pmatrix}, \tag{8}$$

$$= e^{-i\phi/2} \begin{pmatrix} \cos\dfrac{\theta}{2} & -i\sin\dfrac{\theta}{2} \\ -i\sin\dfrac{\theta}{2} & \cos\dfrac{\theta}{2} \end{pmatrix} \begin{pmatrix} e^{i\phi/2} & 0 \\ 0 & e^{-i\phi/2} \end{pmatrix}, \tag{9}$$

where

$$\cos\frac{\theta}{2} = \frac{1}{\sqrt{1 + \mu^2}},$$
$$\phi = E\lambda(n, n - 1).$$

If ϕ were real, the matrix M would be unitary.

The solution for $\Psi^{(n)}$ in terms of the initial $\Psi^{(0)}$ is

$$\Psi^{(n)} = M_{n,n-1} M_{n-1,n-2} \cdots M_{1,0} \Psi^{(0)}. \tag{10}$$

Evaluating the matrix product in (10) does not generally yield easily analyzed results. However, there is one special case which can be readily discussed.

Suppose $\lambda_{n,n-1}$ is a constant λ independent of n and $n-1$. Then

$$M_{\nu,\nu-1} \equiv M,$$

and (10) becomes

$$\Psi^{(n)} = M^n \Psi^{(0)}. \tag{11}$$

We now need only to diagonalize M:

$$M\chi_\pm = e^{i(\pm\psi-\phi/2)}\chi_\pm, \tag{12}$$

where

$$\cos\psi = \cos\frac{\theta}{2}\cos\frac{\phi}{2} \tag{13}$$

and

$$\chi^{(+)} = \begin{pmatrix} e^{i\psi} - \cos\dfrac{\theta}{2}e^{-i\phi/2} \\[2mm] -i\sin\dfrac{\theta}{2}e^{i\phi/2} \end{pmatrix}, \quad \chi^{(-)} = \begin{pmatrix} i\sin\dfrac{\theta}{2}e^{-i\phi/2} \\[2mm] e^{i\phi/2}\cos\dfrac{\theta}{2} - e^{-i\psi} \end{pmatrix}. \tag{14}$$

Note that in the limit of weak coupling ($\mu \to 0$, $\theta \to 0$, $\psi \to \phi/2$),

$$\chi^{(+)} \to 2i\sin\frac{\phi}{2}\begin{pmatrix}1\\0\end{pmatrix}, \quad \chi^{(-)} \to 2i\sin\frac{\phi}{2}\begin{pmatrix}0\\1\end{pmatrix}. \tag{15}$$

More generally

$$\Psi^{(0)} = N\left[\left(e^{i\phi/2}\cos\frac{\theta}{2} - e^{-i\psi}\right)\chi_+ + i\sin\frac{\theta}{2}e^{i\phi/2}\chi_-\right], \tag{16}$$

where

$$N^{-1} = -2\left[1 - \cos\frac{\theta}{2}\cos\left(\psi + \frac{\phi}{2}\right)\right]$$

From (11) and (12) we obtain

$$\Psi^{(n)} = N\left[e^{in(\psi-\phi/2)}\left(e^{i\phi/2}\cos\frac{\theta}{2} - e^{-i\psi}\right)\chi_+ + ie^{-in(\psi+\phi/2)}\sin\frac{\theta}{2}e^{i\phi/2}\chi_-\right]. \tag{17}$$

We discuss two limiting cases.

(a) Weak Coupling ($\theta \to 0$)
Under these circumstances, (13) yields

$$\psi \approx \frac{\phi}{2} + \left(1 - \cos\frac{\theta}{2}\right)\cot\frac{\phi}{2},$$

$$\mathrm{Im}\left(\psi - \frac{\phi}{2}\right) \approx \left(1 - \cos\frac{\theta}{2}\right)\frac{\sinh\left(\Gamma\lambda/2\right)}{\cosh\left(\Gamma\lambda/2\right) - \cos\varepsilon\lambda}. \tag{18}$$

As expected, the amplitude of χ_+ will decrease slowly with each successive "collision" while the rate of decrease of the amplitude χ_- is governed by the lifetime Γ of the excited state.

(b) Strong Coupling ($\cos\theta/2 \to 0$)
Then

$$\psi \approx \frac{\pi}{2} - \cos\frac{\theta}{2}\cos\frac{\phi}{2}$$

and

$$\mathrm{Im}\,\psi \approx -\cos\frac{\theta}{2}\sin\frac{1}{2}\varepsilon\lambda\sinh\frac{\lambda\Gamma}{4}. \tag{19}$$

We see that the ratio of the magnitudes of the amplitudes of χ_+ and χ_- changes very little with each succeeding collision. One can verify that the same is true of the ratio of the magnitudes of the ground and excited states. The system thus maintains its "identity," changing slowly with a rate indicated by (19). The amplitude of the whole system diminishes with each collision at half the rate of decay of the excited state.

The physical reason underlying this result is the high probability of the excited state making a transition to the ground state rather than decaying in virtue of the strong coupling.

Acknowledgment

This work was supported in part through funds provided by the U.S. Department of Energy under contract DE-ACO2-76ERO3069.

Reference

[1] For a brief summary see H. Feshbach, *Prog. Particle and Nucl. Phys.* 4: 451 (1980).

How to Analyze Low-Energy Scattering

Robert L. Jaffe

A few years ago Francis Low and I published a paper on low energy hadron-hadron scattering in models with confined quarks and gluons [1]. In it we suggested a new and unconventional way for experimentalists to analyze their data if they want to test the spectroscopic predictions of quark models. As we described it, our scheme appears to require the entire S matrix—phases and elasticities in all open channels—to make a comparison with theory possible. Unfortunately, complete data are not available except for the simplest cases. Even in the $I = 0$, $J^{PC} = 0^{++}$ channel of meson-meson scattering, we are limited to $\sqrt{s} < 2M_\eta \approx 1100$ MeV because there is no data on $\pi\pi \to \eta\eta$.

Typically, experimentalists deal with incomplete data by fitting the predictions of a model to whatever data are available. For reasons that were not altogether clear to us at the time, this method did not seem to suit our model. In the past two years several people have tried to fit data with our model, found it difficult, and emphasized to me how much this limits its usefulness [2]. Happily I now understand the problem and would like to present Francis with the solution and some words about its implications for our picture of low-energy scattering on the occasion of his sixtieth birthday. At the same time I hope this paper will serve as a "primer" for those who would like to use our techniques to analyze low-energy hadron-hadron scattering. With this in mind I will give a brief summary of [1] before raising the problem and describing its resolution. Francis and anyone else familiar with [1] may wish to skip to section II.

I. The P Matrix

In our paper Francis and I proposed a radical change in the way one should think about and analyze low-energy scattering. We showed that the Hamiltonian eigenstates of confined-quark models need not correspond to resonances, so it is not necessarily proper to compare their spectroscopic predictions directly with tables of resonances. Instead we proposed a new way to analyze hadron-hadron scattering designed specifically to extract the masses and coupling constants calculated in quark models.

Boundary conditions are crucial. With scattering-state boundary conditions—allowing incoming and outgoing waves in open channels—eigenstates are indentified with poles in the S matrix and, if near enough to the physical region, appear as resonances. In quark-model spectroscopy, however, the coupling to open channels is usually ignored. In the bag model, for instance, s-wave configurations of arbitrary numbers of quarks and antiquarks (in a color singlet) are constructed by populating the lowest mode in a spherical cavity. The states obtained this way are not poles in S. They obey the wrong, universally confining, boundary conditions. What Francis and I did was to find a quantity, the P matrix, whose poles correspond to the eigenstates of an internal Hamiltonian subject to confining boundary conditions in all channels. P is algebraically related to the S matrix and can be reconstructed from scattering data: by measuring phases and elasticities one may construct P, look for poles, and compare their masses and residues with the predictions of confined-quark models.

The skeptic may well ask why such sophistication is necessary. Theorists have been identifying quark-model eigenstates with resonances since time immemorial without haggling about boundary conditions. The reason the P matrix has come along so late in the history of the quark model is that until recently we have been studying and identifying only $Q\bar{Q}$ and Q^3 configurations. These systems couple to open channels only after creating a quark-antiquark pair. Since quark creation is suppressed in QCD (viz., the OZI rule), open channels are not important and universally confining boundary conditions are a good approximation. Physically we identify the confining boundary condition with the strong color-confining forces which develop when one quark begins to separate from the other(s). In the end we study $Q\bar{Q}$ and Q^3 in a zero-width approximation and believe the effect of open channels is only to give them a small width.

This argument fails for multiquark configurations. There are both confined and unconfined channels. Consider $Q^2\bar{Q}^2$—as I will for the purpose of illustration throughout this article—the $(Q\bar{Q})^8$–$(Q\bar{Q})^8$ channel is confined but the $(Q\bar{Q})^1$–$(Q\bar{Q})^1$ is not (the **1** and **8** are color labels). The admixture of confined and unconfined components in a $Q^2\bar{Q}^2$ system is determined not by hand, but by the color-dependent forces inside the hadron. In the absence of some other barrier, the $Q^2\bar{Q}^2$ system will couple more intimately to open channels than ordinary mesons and baryons do. Yet in models the presence of open channels is usually ignored: in the spherical bag model the quarks are confined by cavity boundary conditions irrespective of their color configuration. The cavity boundary condition is inappropriate in the $(Q\bar{Q})^1-(Q\bar{Q})^1$ channel, and must be regarded as an artificial construct erected for calculational purposes. Quark-gluon eigenstates subject to universally confining boundary conditions should not be thought of as "states" in the usual sense. They may or may not resonate. For this reason Francis and I called them by a new name: "primitives." The skeptic may again be heard asking why not simply use the correct scattering-state boundary conditions? Of course we would, if we knew how to handle the fission and fusion of bags full of relativistic, interacting quarks. We don't, and instead have chosen to find ways to make use of information obtained with the physically unrealistic, but technically convenient, confining boundary condition.

The object of the P-matrix formalism is not merely to discover which quark-gluon primitives resonate. We are also able to associate primitives with entirely nonresonant amplitudes, which ordinarily would be labeled "background" and not considered of interest. QCD is tested by comparing the masses and residues of P-matrix poles in the data with theory regardless of the resonant character of the amplitude associated with the P matrix pole. In [1] we found that such diverse phenomena as the low-energy enhancement in the $I = 0$ $\pi\pi$ s wave and the weakly repulsive phase of the $I = 2$ $\pi\pi$ s wave were associated with P-matrix poles in good agreement with the predictions of the quark bag dynamics.

Formally, the P matrix is defined in the channel space of the two-body scattering problem. Its poles occur at energies $E(b)$ obeying

$$H\psi = E(b)\psi \tag{1}$$

and

$$\psi(\mathrm{r})\big|_{r=b} = 0, \tag{2}$$

where E is the center-of-mass energy $E(b) = \sqrt{s(b)}$ and $\psi(\mathrm{r})$ is the two-body wave function in the relative coordinate r. In [1] Francis and I show that the poles in P factorize, i.e., their location is independent of the channel in which they are observed. Because of this we can think of primitives as intrinsic to the quark-gluon system. We also showed that the residues of poles in P are positive and given by $-\partial s(b)/\partial b$. We gave a prescription for estimating residues in quark models. We associate the boundary condition $\psi(b) = 0$ on the hadron-hadron wavefunction with universally confining bag boundary conditions on the quarks and gluons. This is a crucial (and controversial) step. Confinement to a sphere of radius R is not obviously equivalent to the condition $\psi = 0$ at some $b = b(R)$. The case for making this leap and a form for $b(R)$ are presented in [1] and discussed further in [3] and [4]. In this article I will set $b = 7$ GeV^{-1} (≈ 1.4 fm.) which is a good approximation to the result of [1].

Starting from (1) and (2) it is straightforward to relate the P matrix to the S matrix for any number of two-body channels with arbitrary angular momenta. Hadrons may continue to interact by conventional exchanges when separated by $r > b$. If exchanges are important they must (and can) be included. Francis and I found that they could be ignored in the cases of interest to us. In this event the relation between P and S is quite simple. For the multichannel s wave,

$$S = -e^{-ikb} \frac{1 - \dfrac{i}{\sqrt{k}} P(s, b) \dfrac{1}{\sqrt{k}}}{1 + \dfrac{i}{\sqrt{k}} P(s, b) \dfrac{1}{\sqrt{k}}} e^{-ikb}, \tag{3}$$

where P depends both on the center-of-mass energy and on b through the boundary condition of (2), and k is the matrix of center-of-mass momenta diagonal in the channel space. For higher partial waves (3) generalizes to

$$S = (e^{+\prime}(kb) - \bar{P}e^{+}(kb))^{-1}(e^{-\prime}(kb) - \bar{P}e^{-}(kb)), \tag{4}$$

where $e_l^{\pm}(x)$ and $e_l^{\pm\prime}(x)$ are again diagonal matrices in the channel space. They depend on the channel orbital angular momentum l, and are related to spherical Hänkel functions:

$$e_l^{+}(x) = i(-1)^l x h_l^{(1)}(x), \tag{5}$$

$$e_l^{-}(x) = -i(-1)^l x h_l^{(2)}(x). \tag{6}$$

For $l = 0$, $e_l^{\pm}(x) = e^{\pm ix}$; for $l = 1$, $e_l(x) = i(1 \pm i/x)e^{\pm ix}$, etc. In (4),

$$\bar{P} \equiv \frac{1}{\sqrt{k}} P \frac{1}{\sqrt{k}}.$$

In the case of a single channel with angular momentum zero the relation between P and S is particularly simple. Defining the phase shift δ_0 by $S = e^{2i\delta_0}$, it follows from (3) that

$$P(s, b) = k \cot(kb + \delta_0). \tag{7}$$

At first glance it looks like there must be some mistake. Even for a non-interacting system, $\delta_0 = 0$, $P(s, b)$ possesses an infinite number of poles

$$P_0(s, b) = k \cot kb = \frac{1}{b}\left(1 + 2k^2 \sum_{n=1}^{\infty} \frac{1}{k^2 - (n^2\pi^2/b^2)}\right). \tag{8}$$

Can this be right? Indeed it is, and understanding why is very enlightening. We must consider this ∞ of poles from both the interior and exterior perspectives. First from the exterior: in the case of no interaction the hadron-hadron s-wave scattering wavefunction is

$$\psi(r) = A \sin kr/kr; \tag{9}$$

this has a zero at $r = b$ whenever $kb = n\pi$, hence the poles in P. Viewed from the interior, the noninteracting hadron-hadron system is still a quark-gluon configuration. If we subject it to universally confining boundary conditions we cannot help but generate an infinite number of primitives. The masses and residues will conspire to build up the function $k \cot kb$. This way the dynamics with universal confinement builds up the noninteracting two-body continuum. Francis and I called this phenomenon "compensation," and the poles in $P_0(s, b)$ "compensation poles." They will figure importantly in the rest of this paper.

II. The Problem

Let me analyze some simple data using P-matrix methods. Consider the $I = 2$ s-wave $\pi\pi$ phase shift shown in figure 1. Equation (7) applies, so I can construct $P_{\text{expt}}^{I=2}(s, b)$, shown in figure 2 for $b = 7\,\text{GeV}^{-1}$. It clearly has a pole at $\sqrt{s_0} \approx 1.05\,\text{GeV}$. The residue is harder to extract. It is approximately 0.25–0.30 GeV^3. Such is the analysis of [1].

Suppose I try to invert the process and fit a P matrix to the data. In the case at hand one might wish to see if the single P-matrix pole provides a good description of the data over the whole energy range. For more complicated problems where complete data do not exist, it is impossible to invert (4) so one *must* resort to fitting the P matrix to the data. Since $P_{\text{expt}}^{I=2}$ has a single pole in the energy region of interest it seems natural to attempt to parametrize P with

$$P_{\text{fit}}^{I=2}(s, b) = c + \frac{\lambda_0}{s - s_0}. \tag{10}$$

I have added a constant c, recognizing that even the trivial P matrix requires subtraction (note the $1/b$ in (8)). The dashed curve in figure 3

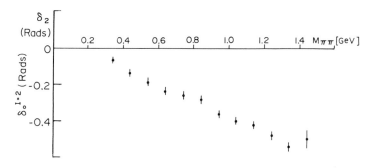

Figure 1
$\pi\pi$ isospin-2 s-wave phase shift.

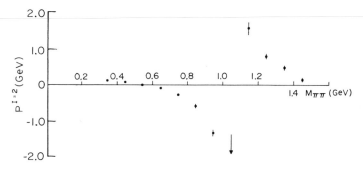

Figure 2
P matrix obtained from the data of figure 1.

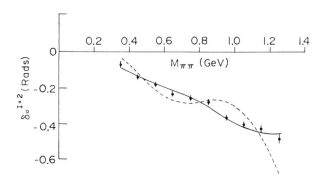

Figure 3
P matrix fits to the data of figure 1: dashed line fit to eq. (10), solid line to eq. (13).

shows the result of fitting (10) to the $I = 2$ $\pi\pi$ s-wave phase shift. The fit is terrible. The pole is at $\sqrt{s_0} = 1.030$ GeV with residue $\lambda_0 = 0.374$ GeV^3, but the χ^2 is 84 for 7 degrees of freedom!

A clue to what has gone wrong can be found by looking at a more traditional scheme for parameterizing a unitary two-body S matrix. Consider the K matrix, defined by

$$S = \frac{1 - ikK}{1 + ikK},\tag{11}$$

in the s wave. For one channel.

$$K = \frac{1}{k}\tan\delta.\tag{12}$$

Notice that $K = 0$ implies $\delta = 0$ and vice versa. A local variation in the phase such as a resonance can therefore be associated with a local effect in K such as a pole. If confined quark model eigenstates corresponded to K-matrix poles (which they do not!) we could start with the trivial K matrix $K_0 = 0$, and add poles to it according to the dictates of our model, resting assured that the S matrix will deviate from unity only in the vicinity of the poles we have introduced.

In contrast, $P = 0$ corresponds to the phase shift $\delta = (\pi/2) - kb$—a highly nontrivial and unphysical starting point for fitting data. Indeed, looking back at figure 3, it appears that except for the wiggle introduced by the pole, the dashed curve is struggling to look like $d\delta/dk = -b$.

III. The Solution

The way around this problem is now obvious: $P = 0$ is the wrong starting point. One should begin with the "trivial" P matrix $P_0(k, b) = k\cot kb$ corresponding to $\delta = 0$. The poles in P_0 are to be identified with the infinite tower of $Q^2\bar{Q}^2$ primitives required in the noninteracting meson-meson system. To fit data in the P-matrix scheme one should *move the poles in P_0 and adjust their residues as required*, interpreting them as $Q^2\bar{Q}^2$ primitives, and one should *add poles to P_0 as required*, interpreting them as $Q\bar{Q}$ (or glueball!) primitives since these are not present at all in P_0. To test this out I return to the $\pi\pi$ $I = 2$ data. Only one $Q^2\bar{Q}^2$ primitive is expected in the energy region of interest and $Q\bar{Q}$ primitives are not allowed. The new algorithm suggests a fit of the form

$$P_{\text{fit}}^{I=2}(s, b) = c + \frac{\lambda_0}{s - s_0} + \left(k\cot kb - \frac{2\pi^2/b^3}{k^2 - (\pi^2/b^2)}\right).\tag{13}$$

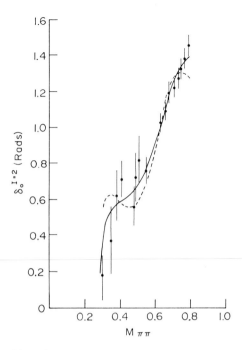

Figure 4

P matrix fits to the $\pi\pi$ isospin-0 s-wave phase shift below 800 MeV: dashed line fit to eq. (10), solid line to eq. (13).

I have "freed" the first pole in P_0, letting its residue and location be fitting parameters. The new fit is shown as the solid line in figure 3. The pole is at $\sqrt{s_0} = 1.054$ GeV with residue $\lambda_0 = 0.334$ GeV and the χ^2 is 6.3 for 7 degrees of freedom, a great improvement.

To check that the improvement is not accidental, I have repeated the analysis for the $I = 0$ $\pi\pi$ s wave. We know there to be a P-matrix pole in the vicinity of 650 MeV from the analysis of [1]. The new fitting algorithm confirms this ($\sqrt{s_0} = 0.651$ GeV, $\lambda_0 = 0.054$ GeV^3), and once again gives a superior fit ($\chi^2 = 8$ versus 26 for 12 degrees of freedom). The two fits, analogous to (10) and (13), are compared in figure 4.

The fitting algorithm is easily generalized to a problem with N channels, and to higher partial waves. Define the "compensating P matrix" P_0, to be the P matrix corresponding to $S = 1$. For the multichannel s wave, (3) gives

$$[P_0(s, b)]_{\alpha\beta} = \delta_{\alpha\beta} k_\alpha \cot k_\alpha b \tag{14}$$

$$= \sum_{\gamma=1}^{N} \delta_{\alpha\gamma}\delta_{\gamma\beta}\left[\frac{1}{b} + \frac{2k_\gamma^2}{b}\sum_{m=1}^{\infty}\frac{1}{k_\gamma^2 - (m^2\pi^2/b^2)}\right]. \tag{15}$$

In (15) I have displayed the compensation poles explicitly. For higher partial waves, (4) gives

$$[P_0(s, b)]_{\alpha\beta} = \delta_{\alpha\beta}\left(k_\alpha\frac{j_{l_\alpha-1}(k_\alpha b)}{j_{l_\alpha}(k_\alpha b)} - \frac{l_\alpha}{b}\right) \tag{16}$$

$$= \sum_{\gamma=1}^{N} \delta_{\alpha\gamma}\delta_{\gamma\beta}\left[\frac{l_\gamma+1}{b} + \frac{2k_\gamma^2}{b}\sum_{m=1}^{\infty}\frac{1}{k_\gamma^2 - q_{l_\gamma,m}^2}\right]. \tag{17}$$

In (17) the momentum $q_{l_\gamma,m}$ is the mth zero of the spherical Bessel function $j_{l_\gamma}(qb)$. Equation (17) is a pleasant surprise: the generalization to higher partial waves introduces no new complication. Notice that P_0 is diagonal in the channel space as expected.

I would like to interpret the compensation poles in $P_0(s, b)$ as primitives of the internal quark-gluon system whose locations and residues conspire to build up the noninteracting two-body continuum. This places several restrictions on the form of $P_0(s, b)$ that must be checked: first the poles in $P_0(s, b)$ must factorize, and second, they must appear at physical values of s. Factorization is obvious from the forms of (15) and (17). The mth compensation pole in the γth channel has a residue of the form $\xi_\alpha^{(\gamma,m)}\xi_\beta^{(\gamma,m)}\lambda(\gamma, m)$ (with $\xi_\alpha^{(\gamma,m)} = \delta_{\alpha\gamma}$ and $\lambda(\gamma, m) = 2q_{l_\gamma,m}^2/b$), as required by factorization.

The restriction that compensation poles appear at physical values of s is more problematic. They emerge naturally as poles in k^2. If the scattering particles in the γth channel have the same mass M_γ, then the associated poles in s occur in the physical region. A compensation pole in the γth channel at q_γ^2 gives rise to a pole in s at $s_\gamma = 4(q_\gamma^2 + M_\gamma^2)$. Since the values of q_γ^2 are always real and positive, the s_γ are always real and above threshold. If, on the other hand, the scattering particles in the γth channel have different masses M_γ and m_γ, a single pole in k^2 maps into two poles in s:

$$\frac{1}{k^2 - q_\gamma^2} = \frac{4s_\gamma^+}{(s_\gamma^+ - s_\gamma^-)(s - s_\gamma^+)} - \frac{4s_\gamma^-}{(s_\gamma^+ - s_\gamma^-)(s - s_\gamma^-)} \tag{18}$$

at

$$s_\gamma^{\pm} \equiv (\sqrt{q_\gamma^2 + M_\gamma^2} \pm \sqrt{q_\gamma^2 + m_\gamma^2})^2. \tag{19}$$

Note that s_γ^- is unphysical: $0 < s_\gamma^- < (M_\gamma + m_\gamma)^2$, and vanishes like $1/q_\gamma^2$ as $q_\gamma \to \infty$. The pole at s_γ^- has no simple interpretation. It is below threshold, so it cannot be thought of as a piece of the "confined continuum."

Furthermore the residue is negative. Needless to say, however, poles in P at unphysical values of s with negative residues do not contradict any deep truths (e.g., unitarity, analyticity), since they occur even when $S = 1$. I have no entirely satisfactory solution to this problem, but I do have a practical one: *simply disregard the poles at* s_γ^-. So I propose the following modification of (17):

$$[P_0(s, b)]_{\alpha\beta} \to \sum_{\gamma=1}^{N} \delta_{\alpha\gamma}\delta_{\beta\gamma}\left[\frac{l_\gamma + 1}{b} + \frac{8k_\gamma^2}{b}\sum_{m=1}^{\infty}\frac{s_{\gamma,m}^+/(s_{\gamma,m}^+ - s_{\gamma,m}^-)}{s - s_{\gamma,m}^+}\right], \qquad (20)$$

where $s_{\gamma,m}^+$ is obtained by setting $q_\gamma = q_{l_\gamma,m}$ in (19). Equation (20) is exact when $m_\gamma = M_\gamma$. It becomes a poorer approximation as $(M_\gamma^2 - m_\gamma^2)/q_\gamma^2$ becomes appreciable. The point of the compensating P matrix is to build up the trivial S matrix. The test of whether the unphysical poles can be dropped is therefore whether the resulting P matrix generates a good approximation to $S = 1$. *For all cases of interest it is an excellent approximation.* For example, in the πK s wave, the absolute value of the phase shift produced by (20) is less than 0.01 radian for $0 < k < 2$ GeV! For the πN s wave, despite the fact that $M_N \gg m_\pi$, the absolute value of the phase shift produced by eq. (20) is less than 0.05 radian. In higher partial waves the minimum value of q_γ is larger and the approximation is better. In short, for practical purposes, the effect of ignoring the poles at unphysical values of s_γ is entirely negligible. From now on I will take (20) as the defining equation for the compensating P matrix.

In particle physics we are used to associate poles with some striking and important effect. Here I have introduced an infinite tower of poles whose sole purpose is to cancel the factor e^{-2ikb} (as in (3)). The phase and the infinite tower of poles appear when we divide the problem into interior and exterior regions at b. We must live with them until we have a unified description of the dynamics at small and large distances. Given the boundary at b it is automatic that $P(s, b)$ will have an infinite tower of poles in every channel. Asymptotic freedom suggests that at high energies the P-matrix poles approach their compensation masses. It is therefore quite natural to hope for a description of low-energy data in which a few low-energy poles are moved from their compensation locations by QCD interaction but the rest remain fixed. Looking at figures 3 and 4 it may still be difficult to accept the necessity of all of these higher poles. The description of low-energy data should not, one feels, depend on what one assumes about the structure of the P matrix at higher energies. In reply to this anxiety I must counter that in more conventional fitting schemes (e.g., K-matrix fits) one prejudices the low-energy behavior equally by choosing not to add arbitrary higher poles. It is no more arbitrary to introduce the compensation poles in the P-matrix fitting algorithm than to leave out arbitrary higher poles in the K-matrix scheme!

Finally I would like to comment on the constants which appear in addition to poles in the P matrix. The P matrix in general needs subtraction:

$$\lim_{|s|\to\infty} P_{ij}(s, b) = K_{ij},$$

$\arg s \neq 0$

where K_{ij} is a constant matrix. This may be checked explicitly in the absence of interactions, and there is no reason to except interactions to change the result qualitatively. Our model only predicts the locations and residues of poles in P, but not the constants K_{ij}. Therefore allowance must be made for an arbitrary constant matrix in P-matrix fits. This certainly detracts from the power of the analysis, but I know of no way around it.

IV. Some Examples

To illustrate the P-matrix scheme I would like to present some more examples of fits to meson-meson scattering. In some cases I have already carried out the fits and will present the results. I will describe one more complicated program which has yet to be attempted. Before attempting a fit it is necessary to determine which channels to include, and how to parameterize the P matrix in each channel. In principle all open channels must be included. In practice the data may suggest that some channels can be ignored. For example in $\pi\pi$ scattering the p wave remains essentially elastic at masses well above $K\bar{K}$ and $\eta\eta$ thresholds. It can be approximated as a one-channel problem. Once the channel space is chosen it is necessary to parameterize the P matrix. Again one should be guided by the data. All the compensation poles in the range spanned by the data must be freed and fit. Data which shows clear and unambiguous resonant behavior will likely require poles to be *added* to the compensation P matrix. I do not know how to make this rule more precise. It is based on an argument which, strictly speaking, does not apply to physical hadron-hadron scattering: Suppose one considers a one-channel problem (to be concrete I will choose $l = 0$ though any l would do) and attempts to describe the phase shift by freeing and fitting some finite number of poles N:

$$P(s, b) = c + \sum_{m=1}^{N} \frac{\lambda_m}{k^2 - k_m^2} + \left(k \cot kb - \sum_{m=1}^{N} \frac{2m^2 \pi^2 / b^3}{k^2 - (m^2 \pi^2 / b^2)} \right). \qquad (21)$$

For k large enough the phase shift $\delta(k)$ returns to zero. (This is true even in the presence of the constant c). It is not possible to describe any *net* resonant behavior merely by moving around poles in the compensating P matrix. This result applies equally well to the multichannel case: as $k \to \infty$ all phases and elasticities return to their trivial values: $\delta_\gamma = 0$, $\eta_\gamma = 1$. This fails to apply to hadron-hadron scattering because the P-matrix formalism breaks down, due to the opening of multibody channels, long before any $k \to \infty$ argument can be applied. A single narrow resonance could be described arbitrarily well by freeing the first compensation pole to fit the rapid phase variation and moving the next N poles down one notch. The price for this perversity would not be paid until $k \approx N\pi/b$; or, in practical terms, for N large enough, never! In practice this means we cannot determine the quark content of a primitive from the form of a P matrix alone: resonant behavior in meson-meson scattering could be attributed either to an added pole (to be identified with a predominantly $Q\bar{Q}$ or glueball primitive) or a shifted compensation pole (to be identified with a predominantly $Q^2\bar{Q}^2$ primitive), provided higher primitives also shift downward to prevent the amplitude rapidly undoing its resonant motion. I regard the latter conspiracies to be possible but unlikely, and am therefore inclined to identify added poles with $Q\bar{Q}$ or glueball primitives. The true test, of course, is the comparison of P-matrix masses and residues with the masses and channel couplings computed in quark models.

With these considerations in mind I will discuss some examples:

1. The $\pi\pi$ p Wave ($J^{PC} I^G = 1^{--} 1^+$)
What primitives are expected? Obviously the ρ, the $Q\bar{Q}$ s wave expected around 750–800 MeV. The first $Q^2\bar{Q}^2$ primitive is expected at a much greater mass. To obtain negative parity it would have to be an orbital excitation costing hundreds of MeV in kinetic energy and lost color-spin attraction (color spin forces are strongest in the s wave).[1] With no $Q^2\bar{Q}^2$ primitive expected it would be an embarassment if a compensation pole fell at low energy. If it did we would have to fit its location and residue

---

1. Negative parity allows the $Q\bar{Q}$ ground state but requires an excitation in $\bar{Q}^2 Q^2$. In contrast positive-parity channels like 0^{++} couple to the $\bar{Q}^2 Q^2$ ground state but only to excitations of $\bar{Q}Q$. This accounts in part for the importance of $\bar{Q}^2 Q^2$ configurations in the meson-meson s wave.

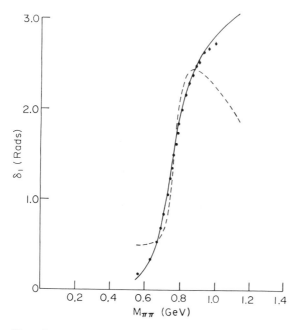

Figure 5

P matrix fits to the $\pi\pi$ isospin-1 p-wave phase shift below 1 GeV: dashed line fit to a single pole plus a constant; the solid line fit to eq. (22).

and interpret it as a $Q^2\bar{Q}^2$ primitive. Fortunately none does. The first compensation pole in the p wave occurs at $q_1 = 4.49/b$ or at $\sqrt{s_1} = 1.31$ GeV in the $\pi\pi$ channel (for $b = 7 \text{ GeV}^{-1}$).

So for $\sqrt{s} \lesssim 1$ GeV a P matrix of the form

$$P(s, b) = c + \frac{\lambda_\rho}{s - s_\rho} + \left(\frac{kj_0(kb)}{j_1(kb)} - \frac{1}{b} \right) \tag{22}$$

should do, where the final two terms are the compensating P matrix for $l = 1$. The fit for $\sqrt{s} < 1$ GeV is shown in figure 5 along with a fit which neglects the compensating P matrix. The ρ pole appears at $\sqrt{s_\rho} = 0.789$ GeV with residue $\lambda_\rho = 0.0375 \text{ GeV}^3$. The constant c turns out consistent with zero. The shift in $\sqrt{s_\rho}$ from the mass of the ρ quoted in the particle data tables is significant. It represents the effect on the ρ mass of eliminating its coupling to the open $\pi\pi$ channel, and has important applications for quark models. Calculations which ignore coupling to decay channels are describing P-matrix poles. Thus a calculation of isospin violation in the $\rho\omega$ system which treats both ρ and ω as $Q\bar{Q}$ should not use the physical ρ and ω masses but instead the locations of the associated

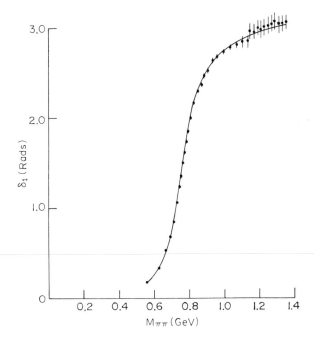

Figure 6
P matrix fit to the $\pi\pi$ isospin-1 *p*-wave phase shift below 1.35 GeV; fit to eq. (23).

P-matrix poles. Dosch has recently studied similar shifts in the charmonium sector [5]. Next let us extend the fit above 1 GeV, ignoring the coupling to $K\bar{K}$. Analyses of $\pi\pi$ scattering data suggest this is a good approximation for $\sqrt{s} \lesssim 1.35$ GeV [6]. At the very least the first compensation pole must be "freed" and fit. It cannot be left at $\sqrt{s_1} = 1.31$ GeV, because being there it requires $\delta_1^1 = n\pi$ at $s = s_1$. So a fit with the ρ and one $Q^2\bar{Q}^2$ primitive is of the form

$$P(s, b) = c + \frac{\lambda_\rho}{s - s_\rho} + \frac{\lambda'}{s - s'} + \left(\frac{kj_0(kb)}{j_1(kb)} - \frac{1}{b} - \frac{2q_1^2/b}{k^2 - q_1^2}\right), \tag{23}$$

where $q_1 b = 4.49$. The data and fit for $\sqrt{s} < 1.35$ GeV are compared in figure 6. Since the phase shift is relatively flat and nearly equal to π for $\sqrt{s} \gtrsim 1$ GeV, it is not surprising that the $Q^2\bar{Q}^2$ primitive is a little shifted from its compensation mass and residue: $\sqrt{s'} = 1.35$ GeV and $\lambda' = 0.442$ GeV^3 (to be compared with compensation values of 1.31 GeV and 0.470 GeV^3, respectively). Including the second pole shifts the ρ mass to $\sqrt{s_\rho} = 0.788$ GeV. Once again the constant c is small and may be set to zero. So the P matrix provides an excellent 4-parameter description of the data below 1.35 GeV, in which all of the parameters have physical significance. In particular to describe the slow variation in the $\pi\pi$ p-wave phase in the energy range $1.0 < \sqrt{s} < 1.35$ GeV a confined quark model should predict a (presumably $Q^2\bar{Q}^2$) primitive with mass ≈ 1.35 GeV and residue ≈ 0.44 GeV^3 in a confining radius $b = 7$ fm^{-1}!

2. The $K\pi$ s Wave at Low Energy

This system is analogous to the $\pi\pi$ s wave except that $m_K \neq m_\pi$. It possesses an exotic ($I = 3/2$) and a nonexotic ($I = 1/2$) channel. The exotic channel is elastic up to $K^*\rho$ threshold. It can be described in terms of a single $Q^2\bar{Q}^2$ primitive. The compensating P matrix is given in eq. (20). The lowest compensation energy is 1.14 GeV, well within the energy region of interest. To fit the data, "free" and fit the first pole and leave the rest unchanged. The resulting fit is so similar to figure 1 that I need not show it. The nonexotic channel is expected to show a $Q^2\bar{Q}^2$ primitive at low energy (≈ 950 MeV) [7] and a $Q\bar{Q}$ primitive somewhere above 1 GeV. Since the $K\eta$ threshold lies at 1.04 GeV, it is not possible to fit for both the $Q\bar{Q}$ and $Q^2\bar{Q}^2$ primitives without expanding the channel space to include $K\eta$. The data below, say, 1 GeV can be fit with only the $Q^2\bar{Q}^2$ primitive, and the fit confirms the bag model expectation of a low-mass cryptoexotic $Q^2\bar{Q}^2$ primitive.

3. The $J^{PC}I^G = 0^{++}0^+$ Channel between $\pi\pi$ and $\rho\rho$ Thresholds

This is perhaps the most interesting (because of the potential presence of glueballs), most active, and least understood system in light-quark spectroscopy. Below the $\rho\rho$ threshold the two-body channels are $\pi\pi$, $K\bar{K}$, $\eta\eta$ (I will ignore $\eta\eta'$ here because its threshold is only a few tens of MeV below $\rho\rho$). What primitives are expected? Certainly the $(Q\bar{Q})0^{++}$ nonet provides two primitives in this energy range. If magically mixed

Table 1

Name (see [7])	Quark Content	Predicted Mass (MeV)	Predicted Coupling To[a]				
			$\pi\pi$	K K	$\eta\eta$	$\eta'\eta$	$\eta'\eta'$
$C^0(9)$	$\bar{u}u\bar{d}d$	650	0.87	—	0.24	0.35	0.26
$C^s(9)$	$\bar{s}s(\bar{u}u + \bar{d}d)$	1100	—	0.71	0.48	—	−0.52
$C^0(36)$	$\bar{u}u\bar{d}d$	1150	−0.50	—	0.42	0.61	0.45
$C^0(9^*)$	$\bar{u}u\bar{d}d$	1450	0.87	—	0.24	0.35	0.26
$C^s(36)$	$\bar{s}s(\bar{u}u + \bar{d}d)$	1550	—	0.71	−0.48	—	−0.52

[a] Normalized to unity for each primitive.

they contain $\frac{1}{\sqrt{2}}(\bar{u}u + \bar{d}d)$ and $\bar{s}s$, respectively. The lightest $Q^2\bar{Q}^2$ nonet provides two more with quark content $\bar{u}d\bar{d}u$ and $\frac{1}{\sqrt{2}}(\bar{u}u + \bar{d}d)\bar{s}s$. But a more careful study of the tables of primitives in [7] reveals no less than three more in this energy range, with several more lurking at masses just a bit higher. The names, quark content, predicted masses, and expected couplings of all the four quark 0^{++} primitives with mass less then $2M_\rho$ are given in table 1. If there are excited $Q^2\bar{Q}^2$ 0^{++} primitives these too should be included in the table, but I doubt any of these lie at such low energy. I should point out that the $C^0(9*)$ primitive is expected to couple primarily to vector-vector channels.

The next step is to enumerate the compensation poles which occur in this energy range. Since all channels are s waves, the compensation poles appear when the channel momenta are π/b, $2\pi/b$, $3\pi/b$, etc. In the $\pi\pi$ channel they appear at $\sqrt{s} = 0.94, 1.82, \ldots$ GeV, in the $K\bar{K}$ channel at $\sqrt{s} = 1.34, 2.05, \ldots$ GeV, and in $\eta\eta$ at $1.42, 2.10, \ldots$ GeV. To analyze the $0^{++}0^+$ channel thoroughly one must explore a variety of possible forms for the P matrix. The first attempt might be to fit the data without *adding* any poles to P_0. This is likely to fail since the data show at least one clear resonance, the $\varepsilon(1300)$. A recently reported analysis of $K_s K_s$ data finds another resonant effect near 1.77 GeV, dubbed the $S^{*\prime}(1770)$. To study this effect one must either include vector-vector channels or show that they are neglible. The well-known $S^*(980)$ can most likely be understood as a compensation pole strongly coupled to the $K\bar{K}$ channel but shifted below $K\bar{K}$ threshold; in short, a $K\bar{K}$ bound state decaying to $\pi\pi$.

With this in mind my ansatz for fitting the three-channel $\pi\pi$, $K\bar{K}$, $\eta\eta$ system for $\sqrt{s} \lesssim 1.50$ GeV would be

$$P(s, b)_{\alpha\beta} = C_{\alpha\beta} + \sum_{m=1}^{4} \frac{\lambda_{\alpha\beta}^m}{s - s_0^m} + \sum_{\gamma=1}^{3} \delta_{\alpha\gamma}\delta_{\beta\gamma}\left(k_\gamma \cot k_\gamma b - \frac{2\pi^2/b^3}{k_\gamma^2 - (\pi^2/b^3)}\right). \tag{24}$$

Three compensation poles are freed and fit, one pole is added. In addition the (symmetric) matrix subtraction constant $C_{\alpha\beta}$ must be allowed for. If the data show rapid variation which this fit cannot account for, more poles must be added; if more gentle deviations occur, higher compensation poles must be freed and fit. One technical point: the 3-channel P matrix of (24) is appropriate only above all three thresholds. Below $2M_\eta$

the system should be described by a 2-channel P matrix and below $2M_K$ by a 1-channel P matrix. These are obtained from eq. (24) by computing a "reduced P matrix" as explained in [1]:

$$\bar{P}_{oo} = P_{oo} - P_{oc} \frac{1}{P_{cc} + K_{cc}} P_{co}, \tag{25}$$

where o and c represent open and closed channels, respectively. Here $K_{cc} = -ik_{cc}$ is a real positive diagonal matrix in the channel space. The "reduction" process can generate some nontrivial and phenomenologically important effects. For example a broad pole in the two-channel $\pi\pi$, $K\bar{K}$ P matrix can give rise to a narrow pole in the reduced P matrix. Thus a smooth variation in the S matrix above $K\bar{K}$ threshold can show up as a rapid phase variation below $K\bar{K}$ threshold. This is a likely explanation for the narrow $S^*(980)$.

No doubt the results of such a complicated fit to fragmentary data will not be unique. Instead I except to find a set of alternative interpretations of the data which can only be sorted out with the help of other data, such as $\gamma\gamma \to X$ and $\psi \to \gamma X$.

As the reader has probably realized from my hypothetical tone, I have not attempted a multichannel fit to the low-energy meson-meson s wave. I look forward to seeing such a fit performed. It will provide us with concrete numbers to compare with our quark-gluon spectroscopic calculations.

Acknowledgments

I would like to thank SLAC and Prof. D. W. G. S. Lieth for their hospitality during a visit when this work was begun. I am grateful to J. Donoghue, E. S. Durkin, K. Johnson, and, of course, F. E. Low for their comments and suggestions. This work was supported in part through funds provided by the U.S. Department of Energy under contract DE-ACO2-76ERO3069.

References

[1] R. L. Jaffe and F. E. Low, *Phys. Rev. D* 19: 2105 (1979).

[2] A. D. Martin and R. E. Cutkosky, private communications; A. C. Irving, A. D. Martin, and P. J. Done, Durham Univ. preprint.

[3] F. E. Low, in *Pointlike Structures Inside and Outside Hadrons*, edited by A. Zichichi, Plenum, New York (1982).

[4] R. L. Jaffe in *Quarks, Gluons and Jets*, Proceedings of the XIV Rencontre de Moriond, edited by J. Tran Thanh Van, C. N. R. S., Paris (1979), p. 81.

[5] H. G. Dosch, MIT preprint MIT-CTP-959.

[6] See, for example, A. D. Martin and M. Pennington, *Ann. Phys.* (N.Y.) 114: 1 (1978).

[7] R. L. Jaffe, *Phys. Rev. D* 15: 267, 281 (1976).

Compound Bags and Hadron-Hadron Interactions

Carleton DeTar

Two rather different methods have been used in order to extract information about hadron-hadron interactions at short distances from the MIT bag model: the P-matrix method of Jaffe and Low and the deformation-energy method of DeTar. The differences and relative advantages of these methods are discussed. A P-matrix-inspired improvement to the deformation-energy method is described.

Introduction

The success of the MIT bag model in accounting for the spectrum, magnetic moments, and other static properties of the low-lying baryons and mesons [1] encouraged efforts to use the model to obtain information about the interactions of hadrons at short distance. However, the very features of the model that allowed an elegant and simple description of the hadrons themselves became an obstacle to a complete and straightforward treatment of hadron-hadron scattering. Therefore, it was necessary to resort to indirect methods. Two of these methods are described briefly and compared here: the P-matrix method of Jaffe and Low [2] and the deformation-energy method of DeTar [3]. The P-matrix approach is practical, but restricted in its treatment of the quark interactions themselves. The deformation-energy method is impractical in that it does not make a quantitative prediction for phase shifts, but it is more flexible in its treatment of the quark interactions. A comparison of the two methods suggests a possible improvement in the deformation-energy method.

Brief Description of the Bag Model

The MIT bag model [4] describes hadrons as extended objects containing colored quarks and gluon fields, compressed by the bag or vacuum pressure B. In the static cavity approximation, the surface of the hadron is fixed and the interactions of the fields are treated perturbatively in the color coupling constant α_c[1]. In the absence of interactions, the quark spinor q and gluon electric and magnetic fields $\mathbf{E}^a$ and $\mathbf{B}^a$ obey the free Dirac and free Maxwell equations, respectively, with the boundary conditions

$$-i\gamma \cdot \hat{n}q = q; \quad \hat{n} \cdot \mathbf{E}^a = 0; \quad \hat{n} \times \mathbf{B}^a = 0 \tag{1}$$

at the surface of the bag. The vector $\hat{n}$ is the outward normal to the surface. These equations produce the "free" cavity eigenmodes for the quarks and gluons. The perturbative vacuum is defined with respect to these modes. Perturbation theory then proceeds with the usual quantum-chromodynamic Hamiltonian, but quark and gluon propagation is restricted by (1). The total energy of the hadron includes the energy of the fields and the postulated volume energy:

$$E = E_{\text{field}} + BV, \tag{2}$$

where B is the bag constant and V is the volume. The size and shape of the hadron is determined by minimizing the total energy with respect to the orientation of the surface. The low-lying hadrons are assumed to be essentially spherical. With this model their masses and various properties are calculated to first order in α_c[1].

The static cavity approximation is elegant and simple in its treatment of the relativistic bound-state problem. It shares a number of features of the shell model of nuclear physics, including a number of defects: the c.m. momentum is not quantized, and there are spurious excitations of the c.m. coordinate. For the low-lying hadrons these difficulties can be overcome in a reasonably satisfactory way [5, 6]. However, if we wish to treat hadron-hadron scattering and loosely bound hadronic states such as nuclei, the time-honored method of nuclear physics for going beyond the shell model, namely, the generator coordinate method [7], fails to apply. The problem is that the Hamiltonian of the static cavity approximation is defined only with respect to a particular cavity. Thus there is no meaning to a quantum-mechanical superposition of two states with different cavity shapes. Scattering problems can be treated in principle in the original Lagrangian formulation of the MIT bag model [4]. However, no quantized version of the model has been found in more than one space and one time dimension. Therefore, it is necessary to proceed indirectly to a treatment of hadron-hadron interactions.

P-Matrix Method

To lowest order in α_c, hadron-hadron interactions at short range involve the formation of a "compound" bag containing the constituents of the interacting hadrons. The bag model provides a crude description of the spatial organization of the quarks and gluons of the compound state and an estimate of its energy. Generally speaking, such compound bags are unstable against dissociation into the original or related scattering channels. Thus to define them in the static cavity approximation entails the introduction of a constraint. Here the P-matrix and the deformation-energy approaches part company. In the typical P-matrix approach the interacting quarks of the compound bag are constrained to occupy the lowest cavity orbital with $J^P = \frac{1}{2}^+$, called the "$S_{1/2}$ orbital." This constrained state is called a "primitive." In the deformation-energy approach a Lagrange constraint is added to the quark Hamiltonian, which results in a dynamical restriction on the motion of the quarks. However, the constraint may be varied to allow a wider range of configurations to be explored.

The P-matrix approach proceeds by proposing a correspondence between the constraint on the quarks and a constraint in the related non-relativistic two-hadron scattering problem, namely, that the restriction to the $S_{1/2}$ orbital corresponds to forcing the wave function of the two hadrons $\psi(r)$ in the relative coordinate r to have a zero at $r = b$, where b is estimated from the bag size. The zero can be thought of as resulting from trapping the hadrons by adding an infinite barrier to the potential; however, it is also illuminating to regard it as a boundary condition imposed upon the solution to the scattering problem for all r at the same energy as that of the primitive state. Such a boundary condition implies a particular value of the scattering phase shift. The analysis is made straightforward by defining a P matrix, related to the S matrix. Poles of the P matrix occur at those energies for which a node of the scattering solution crosses $r = b$. These energies are the energies of the primitive states.

To be more concrete, let us consider the most elementary case of s-wave scattering in a single channel with no interaction for $r > b$. The two-hadron scattering wave function $\psi(r)$ at c.m. momentum k is given for $r > b$ by

$$\psi(r) = A(k)\left[\cos k(r - b) + \frac{P(k)}{k}\sin k(r - b)\right], \tag{3}$$

which defines the P matrix, or by

$$\psi(r) = N(k)[e^{-ikr} - S(k)e^{ikr}],\tag{4}$$

which defines the S matrix $S(k) = \exp 2i\delta(k)$. Nodes in ψ at $r = b$ are associated with zeros in $A(k)$ and poles in $P(k)$. Equating the two expressions gives the P matrix

$$P(k) = k \cot [kb + \delta(k)].\tag{5}$$

The analysis may be generalized to the case of multiple channels, higher angular momenta, and a residual long-range interaction [2, 8, 9]. A common feature of the analysis remains: an explicit procedure is given for the construction of the P matrix from the scattering data, given a particular value of b and a scattering potential for $r > b$. Poles of the P matrix are associated with the primitive bag states.

To complete the correspondence with the quark state, it is necessary to estimate b. This is done by requiring that the "size" of the primitive be the same, whether described in the bag model or in the potential model. The size in the bag model is taken to be the r.m.s. separation of the clusters of quarks in the bag cavity. For a two-meson primitive, consisting of two quarks and two antiquarks, with all quarks in the $S_{1/2}$ orbital, the r.m.s. separation $\bar{r}$ is given by

$$\bar{r}^2 = \left\langle \left(\frac{\mathbf{r}_1 + \mathbf{r}_2}{2} - \frac{\mathbf{r}_3 + \mathbf{r}_4}{2} \right)^2 \right\rangle = \langle r_i^2 \rangle,\tag{6}$$

where the probability distribution is measured with respect to the baryon density $q^\dagger q$. The size in the potential model is taken to be the r.m.s. separation of two otherwise noninteracting hadrons moving in an infinite square well of radius $r = b$ in the relative coordinate, i.e.,

$$\bar{r}^2 = \int_0^b r^2 \sin^2 (\pi r/b) dr \Big/ \int_0^b \sin^2 (\pi r/b) dr.\tag{7}$$

The result of equating these two separations is $b \simeq 1.4R$, where R is the bag radius.

The P-matrix method has been applied with success to a few meson-meson channels [2]. The energy of the bag primitive is calculated to first order in α_c, using the method of Jaffe [10], keeping only gluon exchange diagrams but neglecting gluon annihilation diagrams, the latter signifying mixing with $Q\bar{Q}$ and $Q^3\bar{Q}^3$ states. Efforts are underway to carry out a similar analysis for two-baryon channels [8, 11].

Deformation-Energy Method

The deformation-energy method introduces a variable constraint and a sufficiently flexible range of configurations that the quark clusters forming the hadrons can be physically separated as the constraint is varied. The energy of the bag configuration is computed and plotted as a function of a measure of cluster separation. The resulting curve is only crudely related to a potential energy of hadron-hadron interaction. The method does not predict a phase shift at any energy. Nevertheless, it gives a qualitative picture of the interaction.

In the simplest version of the method, namely, for a spherical cavity [12], left and right hybrid quark orbitals are defined so that

$$q_L = (q_S - \sqrt{\mu}q_P)/\sqrt{1+\mu},$$
$$q_R = (q_S + \sqrt{\mu}q_P)/\sqrt{1+\mu}, \tag{8}$$

where q_S and q_P are the spinors for the lowest even and odd parity orbitals respectively ($\frac{1}{2}^+$ and $\frac{3}{2}^-$) and μ is a parameter varying from $\mu = 0$, corresponding to complete overlap, to $\mu = 1$, corresponding to complete orthogonality of q_L and q_R. The relative phases of q_S and q_P are chosen so that the shift to the left or right is evident in the densities $q_L^\dagger q_L$ or $q_R^\dagger q_R$. The compound nucleon-nucleon state is formed by placing three quarks in the left orbital with the internal quantum numbers of one nucleon and three quarks in the right orbital in a similar way. The state is then fully antisymmetrized. The resulting state can be characterized for even parity as one of mixed orbital configuration with a component consisting of all six quarks occupying the $S_{1/2}$ orbital and a component with pairs of quarks promoted to the $P_{3/2}$ orbital. A single parameter has been used here in order to define a path through configuration space, although more general parametrizations may be considered [13]. A cluster separation parameter δ is defined so that at small separations

$$\delta \alpha \sqrt{\mu} \int q_P^\dagger q_S z d^3 r, \tag{9}$$

where the left-right axis is the z axis. As δ increases from zero with increasing μ the baryon number density shifts from a spherically symmetric pattern to a bimodal pattern, thereby demonstrating the association between δ and the physical separation of the hadronic clusters.[1] At

---

1. Harvey [13] uses a different definition of separation. Inasmuch as the calculations of [3] and [12] are carried out in a single bag cavity, they correspond in Harvey's language

maximum separation the configuration is, by construction, that of two nucleonic clusters, distorted, of course, by the geometry. More generally, if deformations from a spherical configuration are permitted, increasing δ leads to fission into two nucleons [3].

The constraint is introduced by adding a term to the cavity Hamiltonian:

$$H_{bag}(\lambda) = H_{bag} + \lambda\delta, \tag{10}$$

where λ is a constant. For $\lambda = 0$ there is no constraint, and the configuration relaxes to the ground state of H_{bag} for the spherical cavity. For the two-nucleon configuration and very likely for many two-meson configurations as well, the ground state has $\mu \neq 0$, i.e., a mixed configuration, corresponding to a partial separation of the hadrons. For large positive λ smaller δ are preferred and μ is small. The expectation value of H_{bag} is plotted versus the expectation value of δ to obtain the deformation energy.[2] With this method it was possible to show that the quark model is capable of accounting for the fact that the two-nucleon system has a short-range repulsive core that gives way at intermediate range to attraction [3]. However, no quantitative predictions can be made-concerning scattering phase shifts.

Toward an Improved Deformation-Energy Method

It is traditional to analyze low-energy hadron-hadron scattering in terms of a two-body potential. Clearly a two-body approach has limited validity to the extent that it does not incorporate the larger degree of freedom of the quark model. This drawback can be overcome in part by resorting to a

---

to zero separation. The minimum-energy configuration was found in [3] to be a mixed configuration with a strong $P_{3/2}$-level occupation. The characterization of this configuration as a partially separated nucleon-nucleon system was corroborated by a computation of the distribution of baryon number density, which showed a clear bimodal structure. Although Harvey's calculation uses a more general configuration than that of [3], the relationship between his separation parameter and the physical separation of the quark clusters has not been established. Thus it has not been proven that he is not seeing essentially the same phenomenon as that of [3], but in a different language.

2. This method is procedurally equivalent to the method of Lagrange, in which $H_{bag}(\lambda) = H_{bag} + \lambda(\delta - \delta_0)$ and the ground-state energy is minimized with respect to λ at fixed δ_0 to give $E_{bag}(\delta_0)$ directly. From a computational standpoint the former method is more convenient.

multichannel analysis, including excited hadronic channels, as well as hidden, confined color channels.

The P-matrix method has the advantage that it is not necessary to introduce a phenomenological potential at short range, where the quark degrees of freedom become manifest. One merely sets a quark-model boundary condition upon the external scattering wave function. A potential is used at long range where its validity is more secure. However, the boundary condition provides precise information about the phase shift only at the energies of the primitive states. Moreover, the relationship between the constrained internal quark structure, e.g., all quarks in the $S_{1/2}$ orbital, and a boundary condition in the two-body scattering problem, i.e., infinite well, is at best qualitative; its quantitative meaning in the model must be taken as an article of faith. In an attempt to go beyond the model and, at the same time, to obtain a better understanding of its assumptions, I find it necessary to resort to a potential model at short range.

Consider the most elementary case of a single-channel description with a potential $V(r, p)$, where r is the relative coordinate and p is the relative momentum. The primitive states of such a potential are eigenstates of the Hamiltonian

$$H_P = \frac{p^2}{2\mu} + V(r, p) + V_b(r), \quad r < b, \tag{11}$$

where the constraining potential $V_b(r)$ gives an infinite barrier at $r = b$ and μ is the reduced mass. The phenomenological potential $V(r, p)$ is judged to be "correct," if it produces the same spectrum of primitive states as are found in the quark model in the presence of the *corresponding* constraint. The difficult question is how to formulate the corresponding constraint. To make the correspondence more natural, let us choose, instead, a harmonic-oscillator constraint and write the full Hamiltonian in the form

$$H = \frac{p_1^2}{2m_1} + \frac{p_2^2}{2m_2} + V(r, p) + \lambda(r_1^2 + r_2^2 - 2\mathbf{r_1} \cdot \mathbf{r_2}), \tag{12}$$

where λ is positive. The c.m. motion is easily separated from the internal motion. The ground-state deformation energy $E(\lambda)$ is defined by

$$E(\lambda) = \left\langle \lambda \left| \frac{p^2}{2\mu} + V(r, p) \right| \lambda \right\rangle, \tag{13}$$

where μ is the reduced mass, $|\lambda\rangle$ is the ground state of H, and both the energy of c.m. motion and the contribution of the constraint have been removed from $E(\lambda)$. There is a natural correspondence between this phenomenological two-body Hamiltonian and the constrained bag-model Hamiltonian,

$$H_{\text{bag}}(\lambda) = H_{\text{bag}} + \lambda\left(\frac{1}{N}\sum_{i=1}^{2N} r_i^2 - \frac{1}{N^2}\sum_{i\neq j}\mathbf{r}_i\cdot\mathbf{r}_j\right), \tag{14}$$

where $N = 2$ for mesons and 3 for baryons, and i labels valence quarks and antiquarks. The correspondence is evident if we regard the two hadrons as being two tight clusters each of N constituents at positions $\mathbf{r}_{h1}$ and $\mathbf{r}_{h2}$. Then the constraint term is exactly $\lambda(r_{h1}^2 + r_{h2}^2 - 2\mathbf{r}_{h1}\cdot\mathbf{r}_{h2})$. In a similar manner one can readily associate the more general two-body constraint

$$\lambda(\mathbf{r}_1 - \mathbf{r}_2)^2 + \mu(\mathbf{p}_1 - \mathbf{p}_2)^2 \tag{15}$$

with the quark-model constraint

$$\frac{\lambda}{N}\sum_{i=1}^{2N} r_i^2 - \frac{\lambda}{N^2}\sum_{i\neq j}\mathbf{r}_i\cdot\mathbf{r}_j + \frac{\mu}{N}\sum_{i=1}^{2N} p_i^2 - \frac{\mu}{N^2}\sum_{i\neq j}\mathbf{p}_i\cdot\mathbf{p}_j. \tag{16}$$

A deformation energy of the quark-bag configuration can be defined in analogy with that of the two-body system. The ground state of the constrained Hamiltonian is determined. The ground-state deformation energy $E_{\text{bag}}(\lambda, \mu)$ is defined as usual as the ground-state expectation value of H_{bag} minus the energy of c.m. motion. The goal of the phenomenological analysis is to find a phenomenological potential that produces the same deformation energies as the quark model for a suitable range of the constraint parameters. This range is determined by the limitations of the quark-model analysis. Consider, for example, a quark configuration that is repulsive at short range and dissociates into two mesons at long range. If we choose to deal only with spherical bags, we expect that only for those values of the parameters that force a compression of the quarks is the deformation insensitive to the choice of a spherical bag. It is necessary, of course, to allow distortions of the spherical geometry in order to test the effect of the restriction to spherical shapes. We recall that in the case of the nucleon-nucleon interaction it was found, with a different constraint, that the spherical approximation is good over a substantial portion of the repulsive core [12].

The quark constraint terms are chosen to be elementary, to correspond to a constraint in the relative coordinate, and to respect the identity of the fermions. Indeed for the two-baryon problem the constraint in (14) could be written in terms of the fields in the form

$$\frac{\lambda}{N}\int : q^\dagger q : r^2 d^3 r - \frac{\lambda}{N^2}\int : q^\dagger \mathbf{r} q \cdot q^\dagger \mathbf{r} q : d^3 r. \tag{17}$$

The Pauli principle prevents a precise definition of a relative cluster coordinate for the quark configuration, unless the two clusters are in orthogonal states, as they would be in two separated bags or in the baryon-antibaryon problem. It is desirable nonetheless to choose a variable that best imitates the dynamical coordinate of the two-body problem.

Comparison with the P-Matrix Analysis

The improved deformation-energy method outlined above may be capable of extracting more information from the quark model than the P-matrix approach. Varying the deformation constraint parameter over a wide range is permissible and is rather like varying the matching radius b over a wide range. However, only small variations of b are considered in P-matrix applications. Because the deformation constraint is dynamical, it permits quarks to occupy orbitals other than the $S_{1/2}$ orbital, if such an occupation lowers the energy. It is to be expected that this freedom also provides more information about the quark model than is obtained by restricting the quarks to the $S_{1/2}$ orbital. The improved deformation-energy approach also seeks to establish a clearer correspondence between the constraint in the quark model and the constraint in the two-body system. Its chief drawback, however, is that it requires the introduction of a short range two-body phenomenological potential. A direct link with experimental phase shifts is therefore lost. It remains to be seen whether the new method will prove to be as satisfactory quantitatively as the P-matrix method.

Acknowledgments

I am grateful to Francis Low, Bob Jaffe, and John Negele for comments and criticism. This work was supported in part by the U.S. National Science Foundation, grant number PHY80-08249, and by NORDITA under the auspices of a NORDITA visiting professorship at the Research Institute for Theoretical Physics, University of Helsinki.

References

[1] T. DeGrand, R. L. Jaffe, K. Johnson, and J. Kiskis, *Phys. Rev. D* 12: 2060 (1975).

[2] R. L. Jaffe and F. E. Low, *Phys. Rev. D* 19: 2105 (1979).

[3] C. E. DeTar, *Phys. Rev. D* 17: 302, 323 (1978); 19: 1028(E) (1979).

[4] A. Chodos, R. L. Jaffe, K. Johnson, C. B. Thorn, and V. F. Weisskopf, *Phys. Rev. D* 9, 3471 (1974); A. Chodos, R. L. Jaffe, K. Johnson, and C. B. Thorn, *ibid.* 10: 2599 (1974).

[5] J. Donoghue and K. Johnson, *Phys. Rev. D* 21, 1975 (1980); K. Johnson, in *High Energy Physics in the Einstein Centennial Year, Proceedings of Orbis Scientiae 1979, Coral Gables*, edited by A. Perlmutter, F. Krausz, and L. F. Scott, Plenum, New York (1979), p. 61.

[6] T. A. DeGrand, *Ann. Phys.* (NY) 101: 496 (1976); C. Rebbi, *Phys. Rev. D* 12: 1407 (1975).

[7] D. L. Hill and J. A. Wheeler, *Phys. Rev.* 89: 1102 (1953); J. J. Griffin and J. A. Wheeler, *ibid.* 108: 311 (1957); J. Ribeiro, *Z. Physik C* 5: 27 (1980); M. Oka and K. Yazaki, *Phys. Lett. B* 90: 41 (1980).

[8] R. L. Jaffe and M. Shatz, Cal Tech report CALT-68-775 (1980). R. L. Jaffe in *Proceedings of the 1981 International Symposium on Lepton and Photon Interactions at High Energies*, edited W. Pfeil, Physikalisches Institut, Univ. Bonn, p. 395.

[9] P. J. Mulders, *Phys. Rev. D* 25: 1269 (1982); Nijmegen preprint THEF-NYM-81.6 (1981).

[10] R. L. Jaffe, *Phys. Rev. D* 15: 267, 281 (1977).

[11] R. L. Jaffe, private communication (1981).

[12] C. E. DeTar, *Phys. Rev. D* 24: 752 (1981).

[13] M. Harvey, *Nucl. Phys. A* 352: 326 (1981).

The Chew-Low Theory and the Quark Model

John F. Donoghue

Some of the subtleties associated with the definition of quark and anti-quark number are discussed. This leads to a calculation of the πNN, ρNN, and $\pi N\Delta$ coupling constants, and to some comments on πN scattering in the quark model.

Many contributions of Francis Low's have shaped the way in which we now understand physics. Many of the other contributors to this volume have a deeper understanding of these developments than I, and I must leave it to them to comment on these. Instead I wish to describe some recent work in the quark model on the coupling of pions and nucleons. This is relevant to a celebration of Francis's work, as the Chew-Low model can be viewed as the start of pion-nucleon theory. If one goes back and studies the original Chew-Low paper [1] one cannot fail to be impressed with the power of their work. Theirs was a large advance, and very little has occurred in the nearly thirty years since then. In fact, it appears as if we have retrogressed. We now describe nucleons and pions in terms of quark and gluons, but an adequate description of pion-nucleon scattering appears further from reach now than then.

In addition, the Chew-Low model presents an apparent paradox when combined with the quark model. In the Chew-Low theory the nucleon and the pion are the fundamental particles. By considering their interactions it is seen that a resonance develops in the Δ channel. The existence and, more specifically, location and coupling of the Δ is governed by general considerations such as the form of pion-nucleon coupling and analyticity. In the quark model, however, one always solves the model in the limit of no coupling to real or virtual channels. This procedure yields three fundamental states of interest to us, the nucleon, the pion, and the Δ. The Δ appears to exist quite independently of the pion-nucleon coupling. If one then turns on the coupling to other channels, one would expect to find two Δ states, one the 3-quark Δ and the other the Chew-

Low resonance of π and p. To the extent that the coupling is variable and can be "turned on," these states are independent and we are left with a problem of the number of states. If, as we expect, the two Δ states are really the same state it follows that the coupling cannot be turned off and, from the Chew-Low analysis, should be very important in the study of the Δ. How then can the quark model claim to understand the nucleon and Δ without including the effects of the pion?

This is a specific version of the general question of how simple quark-model ideas can make sense in a strongly interacting world. In the quark model one has a Fock-space decomposition of hadronic states into components with definite numbers of quarks. There are 3-quark baryons, $4q1\bar{q}$ baryons, etc. The ground states are assumed to be dominantly 3 quark, with possibly a *small* admixture of $4q1\bar{q}$. The apparent conflict with the Chew-Low theory would be resolved if the $\pi(= q\bar{q})$ + nucleon $(= 3q)$ resonance was not the $3q$ Δ of the quark model, but a $4q1\bar{q}$ state. However, this is not the correct interpretation; πp should resonate in the $\Delta(1232)$. Somehow the Fock-space decomposition must be modified without destroying the good features of the quark model.

Perhaps a clue is found in the fact that the definition of the number of particles is basis-dependent in a relativistic theory, and we can change the number of quarks simply by changing basis functions, with no real physics occurring (i.e., no pair production). I would like to describe this phenomenon in some detail, and then apply it to the $\pi N\Delta$ system [2].

In field theory we expand the field operator in terms of a complete set of solutions to the equations of motion—wavefunctions—along with creation and annihilation operators. For a fermion operator

$$\psi(x) = \sum_i \{\phi_i^{(+)}(x)b_i + \phi_i^{(-)}(x)d_i^\dagger\}, \tag{1}$$

where $\phi_i^{(+)}$ ($\phi_i^{(-)}$) are the positive (negative) energy solutions of the Dirac equation. If we normalize the wavefunctions such that

$$\int d^3x \phi_i^{(A)\dagger}(x)\phi_j^{(B)} = \delta_{ij}\delta^{AB}, \tag{2}$$

with $A, B = (+)$ or $(-)$, then the canonical commutation rules for ψ imply that

$$\{b_i, b_j^\dagger\} = \{d_i, d_j^\dagger\} = \delta_{ij},$$
$$\{b_i, b_j\} = \{d_i, d_j\} = \{b_i, d_j\} = \{b_i, d_j^\dagger\} = 0. \tag{3}$$

One can use the orthonormality properties of the wavefunctions to invert the field decomposition:

$$b_i = \int d^3x \phi_i^{(+)\dagger}(x)\psi(x),$$

$$d_i^\dagger = \int d^3x \phi_i^{(-)\dagger}(x)\psi(x). \tag{4}$$

We can then use the creation and annihilation operators to create a Fock space. The vacuum is that state which vanishes when operated on by all annihilation operators:

$$\begin{aligned} b_i|0\rangle &= 0 \\ d_i|0\rangle &= 0 \end{aligned} \quad \text{for all } i. \tag{5}$$

The one-particle states are

$$\begin{aligned} |i\rangle &= b_i^\dagger|0\rangle, \\ |\bar{i}\rangle &= d_i^\dagger|0\rangle, \end{aligned} \tag{6}$$

and states with higher numbers of particles are created likewise. The number operator

$$N = \sum_i (b_i^\dagger b_i + d_i^\dagger d_i) \tag{7}$$

counts the particles in a given state.

However, someone else can come along and find a second complete set of solutions and use them to quantize the field operator. Let us label quantities associated with the second set with a tilde underneath, i.e., $\underset{\sim}{\phi}_i^{(+)}$ and $\underset{\sim}{b}_i$. Then we have

$$\psi(x) = \sum_i \{\underset{\sim}{\phi}_i^{(+)}\underset{\sim}{b}_i + \underset{\sim}{\phi}_i^{(-)}\underset{\sim}{d}_i^\dagger\}. \tag{8}$$

The new set of creation and annihilation operators satisfy set of commutation relations identical to (3). Likewise we have inversions

$$\underset{\sim}{b}_i = \int d^3x \underset{\sim}{\phi}_i^{(+)\dagger}\psi(x),$$

$$\underset{\sim}{d}_i^\dagger = \int d^3x \underset{\sim}{\phi}_i^{(-)\dagger}\psi(x); \tag{9}$$

vacuum

$$\begin{aligned} \underset{\sim}{b}_i|\underset{\sim}{0}\rangle &= 0, \\ \underset{\sim}{d}_i|\underset{\sim}{0}\rangle &= 0; \end{aligned} \tag{10}$$

and number operator

$$N = \sum_i (b_i^\dagger b_i + d_i^\dagger d_i) \tag{11}$$

in the new basis.

The inversion relation also tells us how to relate the two sets of operators. We plug an expression for $\psi(x)$, eq. (8), into the inversion (4), to obtain

$$
\begin{aligned}
b_i &= \sum_j (\alpha_{ij} b_j + \beta_{ij} d_j^\dagger), \\
d_i^+ &= \sum_j (\gamma_{ij} d_j^\dagger + \varepsilon_{ij} b_j),
\end{aligned}
\tag{12}
$$

where

$$
\begin{aligned}
\alpha_{ij} &= \int d^3x\, \phi_i^{(+)\dagger}(x)\phi_j^{(+)}(x), \\
\beta_{ij} &= \int d^3x\, \phi_i^{(+)\dagger}(x)\phi_j^{(-)}(x), \\
\gamma_{ij} &= \int d^3x\, \phi_i^{(-)\dagger}(x)\phi_j^{(-)}(x), \\
\varepsilon_{ij} &= \int d^3x\, \phi_i^{(-)\dagger}(x)\phi_j^{(+)}(x).
\end{aligned}
\tag{13}
$$

The reverse transformation is

$$
\begin{aligned}
b_i &= \sum_j (\alpha_{ij}^* b_j + \beta_{ij}^* d_j^\dagger), \\
d_i^\dagger &= \sum_j (\gamma_{ij}^* d_j^\dagger + \varepsilon_{ij}^* b_j).
\end{aligned}
\tag{14}
$$

The commutation relations are form-invariant due to the completeness relations

$$
\begin{aligned}
\sum_j (\alpha_{ij}^* \alpha_{jk} + \beta_{ij}^* \beta_{jk}) &= \delta_{ik}, \\
\sum_j (\alpha_{ij}^* \varepsilon_{jk} + \beta_{ij}^* \gamma_{jk}) &= 0, \\
\sum_j (\gamma_{ij}^* \gamma_{jk} + \varepsilon_{ij}^* \varepsilon_{jk}) &= \delta_{ik}.
\end{aligned}
\tag{15}
$$

These transformations are often called Bogoliubov or canonical transformations. The first term in each (α_{ij} or γ_{ij}) is what we expect from non-relativistic physics; the basis change would simply mix the quark creation operators. However, the last term in each transformation (β_{ij} or ε_{ij}) mixes

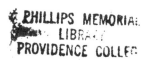

up quark and antiquark degrees of freedom. This plays havoc with our usual ideas about the number of particles. For example, the vacuum of one basis contains particles when looked at in another basis:

$$
\begin{aligned}
\langle \underline{0}|N|\underline{0}\rangle &= \langle \underline{0}|\sum_i (b_i^\dagger b_i + d_i^\dagger d_i)|\underline{0}\rangle \\
&= \langle \underline{0}|\sum_{i,j} \{(\alpha_{ij}^* \underline{b}_j^\dagger + \beta_{ij}^* \underline{d}_j)(\alpha_{ij}\underline{b}_j + \beta_{ij}\underline{d}_j^\dagger) \\
&\qquad\qquad + (\gamma_{ij}\underline{d}_j^\dagger + \varepsilon_{ij}\underline{b}_j)(\gamma_{ij}^*\underline{d}_j + \varepsilon_{ij}^*\underline{b}_j^\dagger)\}|\underline{0}\rangle \quad (16) \\
&= \sum_{i,j}\{|\beta_{ij}|^2 + |\varepsilon_{ij}|^2\}.
\end{aligned}
$$

Likewise the various one, two, three, etc., particle states in one basis contain other numbers of particles in another basis. Of course, when this occurs, the overall states have the same conserved quantum numbers; extra fermions always appear in pairs so as not to change the conserved quantum numbers. However, the point is that the number of quarks is not a conserved quantum number. It will change if we look at the state in a different way (i.e., using the new basis). Also amusing is the fact that if we have a quark plus antiquark where, for some reason, the two particles are defined in different bases, this will sometimes look like no particles:

$$
\langle 0|b_i^\dagger d_j^\dagger|0\rangle = \varepsilon_{ij}^*. \tag{17}
$$

We see that the Fock-space description of the quark model is basis-dependent.

An explicit example may help to make the effect clearer, and will also be useful in later applications. In the MIT bag model one solves the Dirac equation on the interior of the bag (I here use massless quarks):

$$
i\partial\!\!\!/\psi = 0, \tag{18}
$$

with a boundary condition

$$
i\eta\!\!\!/\psi = \psi, \tag{19}
$$

with η^μ being the normal to the surface. In the rigid-cavity approximation the lowest-energy wavefunctions (centered about $\mathbf{x} = 0$) are

$$
\begin{aligned}
\phi_0^{(+)} &= N \begin{pmatrix} ij_0(\omega r)\chi_m \\ -j_1(\omega r)\boldsymbol{\sigma}\cdot\hat{r}\chi_m \end{pmatrix} e^{-i\omega t}, \\
\phi_0^{(-)} &= N \begin{pmatrix} -ij_1(\omega r)\boldsymbol{\sigma}\cdot\hat{r}\chi_m \\ j_0(\omega r)\chi_m \end{pmatrix} e^{+i\omega t},
\end{aligned} \tag{20}
$$

where χ is a two-component Pauli spinor, $\omega = 2.0428/R$, $N = 2.27/\sqrt{4\pi}$ R^3, and R is the radius of the cavity. Solutions for higher-energy modes also exist, forming a complete set for $r < R$.

We can imagine comparing two sets of bagged quarks. One set may be appropriate to a bag located at $x = 0$ with radius R. The other set could be the solutions for a spherical bag at a different point $\mathbf{x} = \boldsymbol{\delta}(t = 0)$, where we will work with $|\boldsymbol{\delta}| \ll R$ for ease. Let us give the shifted bag a radius $R' < R$ such that the whole shifted bag is enclosed within the one located at the origin. Within the smaller bag we have two complete sets, and we can express the operators in the shifted bag in terms of those in the one at $\mathbf{x} = 0$. Forming our overlaps we find, for the ground-state solutions to lowest order in δ,

$$\alpha_{00} = \gamma_{00} = 4\pi NN' \int_0^{R'} dr\, r^2 \{ j_0(\omega r)j_0(\omega' r) + j_1(\omega r)j_1(\omega' r) \},$$

$$\beta_{00} = \varepsilon_{00} = 4\pi NN' \frac{\omega\boldsymbol{\sigma} \cdot \boldsymbol{\delta}}{3} \int_0^{R'} dr\, r^2 \{ j_0(\omega r)j_0(\omega' r) - j_1(\omega r)j_1(\omega' r) \} \quad (21)$$

$$\equiv \bar{\beta} \frac{\boldsymbol{\sigma} \cdot \boldsymbol{\delta}}{R'},$$

with the primed quantities denoting that radius R' is used. If $R' \approx R$ we have

$$\alpha_{00} = \gamma_{00} = 1, \quad (22)$$

$$\beta_{00} = \varepsilon_{00} = \frac{x_0}{6(x_0 - 1)} \frac{\boldsymbol{\sigma} \cdot \boldsymbol{\delta}}{R'} = 0.33 \frac{\boldsymbol{\sigma} \cdot \boldsymbol{\delta}}{R'}.$$

($x_0 \equiv 2.0428 \ldots$). Some cases which are of physical interest are for the nucleon-pion or nucleon-rho systems, where $R = R_N = 5.5$ GeV^{-1}: for $R' = R_\pi = 3.6$ GeV^{-1},

$$\alpha_{00} = \gamma_{00} = 0.58, \quad (23)$$

$$\beta_{00} = \delta_{00} = 0.27 \frac{\boldsymbol{\sigma} \cdot \boldsymbol{\delta}}{R_\pi};$$

or for $R' = R_\rho = 4.4$ GeV^{-1},

$$\alpha_{00} = \gamma_{00} = 0.78, \quad (24)$$

$$\beta_{00} = \varepsilon_{00} = 0.31 \frac{\boldsymbol{\sigma} \cdot \boldsymbol{\delta}}{R_\rho}.$$

This ties up with our previous discussion when we consider the nucleon-pion system. In particular a state composed of a proton at $x = 0$ and a pion at $\mathbf{x} = \boldsymbol{\delta}$, will sometimes look like a 3-quark Δ when expressed in terms of $x = 0$ quarks because the antiquark creation operator in the pion has a piece (proportional to ε_{00}) that removes a quark state

$$\begin{array}{cccc}
qqq & + & qq & \to & qqq \\
x = 0 & & x = \delta & x = 0
\end{array}.$$

The calculation for $P(J_z = \frac{1}{2}) + \pi^+ \leftrightarrow \Delta^{++}(J_z = \frac{1}{2})$ is (not including the effect of spin splitting in the wavefunction of the nucleon and the Δ)

$$\langle P, \mathbf{0}; \pi^+, \boldsymbol{\delta} \,|\, \Delta^{++}, \mathbf{0}\rangle = \frac{2}{3}\alpha_{00}\bar{\beta}\frac{\delta_z}{R_\pi}$$

$$= 0.11\frac{\delta_z}{R_\pi}. \tag{25}$$

This tells us that what we thought was a pure 3-quark state, the Δ, has contained a $4q1\bar{q}$ component all the time, if looked at in a different way. The Bogoliubov transformation brings this out. We see that πP and a bare Δ can be the same without modifying quark-model ideas by explicitly introducing open channels. In this picture the naive quark model can "work" so easily because one has chosen a good basis in which to describe the physics. In other bases the physics, although identical, *looks* more complicated.

Other similar overlaps can be found. For $P\pi^0 \leftrightarrow P$ we have

$$\langle P, \mathbf{0}; \pi^0, \boldsymbol{\delta} \,|\, P, \mathbf{0}\rangle = \frac{5}{3\sqrt{12}}\alpha_{00}\bar{\beta}\frac{\boldsymbol{\sigma}\cdot\boldsymbol{\delta}}{R_\pi}$$

$$= 0.08\frac{\boldsymbol{\sigma}\cdot\boldsymbol{\delta}}{R_\pi}, \tag{26}$$

and for $P(\uparrow) + \rho^0(0) \leftrightarrow P(\uparrow)$ (arrows indicate spin, quantized in the z direction) we have

$$\langle P, \mathbf{0}, \uparrow; \rho^0, \boldsymbol{\delta}, 0 \,|\, P, \mathbf{0}, \uparrow\rangle = \frac{1}{\sqrt{12}}\alpha_{00}\bar{\beta}\frac{\delta_z}{R_\rho}$$

$$= 0.07\frac{\delta_z}{R_\rho}. \tag{27}$$

It appears that these overlaps should have something to do with coupling constants. However, the coupling constants are defined in terms of plane-wave states such as $|N(\mathbf{p})\rangle$, not in terms of the bag-model eigenstates. These are not the same. Ken Johnson and I have shown that bag states may be treated as superpositions of plane-wave eigenstates, with a wavepacket $\phi(\rho)$ or $\chi(\rho)$:

$$|\pi, \mathbf{x}\rangle = \int \frac{d^3 p}{2\omega_p} \phi(p) e^{ip \cdot x} |\pi(p)\rangle,$$

$$|N, \mathbf{x}\rangle = \int d^3 p \frac{m}{E_p} \chi(p) e^{ip \cdot x} |N(p)\rangle. \tag{28}$$

We solved for the pion's wavepacket in the process of calculating the pion decay constant F_π. For heavy states such as the nucleon the precise form of the wave packet is not as important. With this formalism the quark model overlaps are related to some average of the plane-wave states. If we define the pion-nucleon coupling constant by

$$\langle P(p')|\phi_{\pi 0}(x)|P(p)\rangle = \frac{g_{\pi NN}\bar{u}(p')\gamma_5 u(p)}{(p - p')^2 - m_\pi^2} F((p - p')^2) e^{i(p-p') \cdot x}, \tag{29}$$

where $F(q^2)$ is a form factor (called $v(q^2)$ in the Chew-Low paper [1]) normalized to $F(0) = 1$, the overlap in which we are interested becomes

$$\langle P, \mathbf{0}; \pi, \delta|P, \mathbf{0}\rangle = \frac{g_{\pi NN}}{2M_N} \frac{1}{6} \boldsymbol{\sigma} \cdot \boldsymbol{\delta} \int d^3 k \phi(k) F(\mathbf{k}^2) \Big[\int d^3 p \chi^*(p - k)$$
$$\chi(p)(2\pi)^3 \Big] \tag{30}$$

The form factor is essential here as we are not really sensitive to the small-q^2 behavior, as this corresponds to well-separated pions and nucleons. This is just as well, as the quark model could not hope to deal adequately with the pion at small q^2, where its near-Goldstone nature is crucial. Our calculation is sensitive to $q^2 \sim O(1/R_\pi^2)$ dominantly as this is the scale of the wavepacket constructed from $\phi(p)$.

To obtain a numerical estimate let us choose

$$\phi(p) = Ce^{-\alpha p^2}, \tag{31}$$

with

$$\alpha = \frac{(\ln 2)^{1/2} R}{\pi},$$

which is an approximation to the form found by Johnson and myself. C is determined by the normalization conditions on ϕ. A similar form is used for $\chi(p)$. For the form factor we will use

$$F(k^2) = e^{-\langle r^2 \rangle k^2/6}, \tag{32}$$

with $\langle r^2 \rangle \approx \langle r^2 \rangle_{em} = 0.8$ fm^2, which has been chosen to have correctly given the first term in the expansion of $F(k^2)$ (assuming it to behave like the electromagnetic form factors), and also allow the integrals to be done easily. The result of this is

$$\int d^3 k \phi(k) F(k^2) \left[\int d^3 p \chi(p-k)\chi(p)(2\pi)^3 \right] = \frac{0.61}{R_N^2}, \tag{33}$$

which would predict a value for the pion-nucleon coupling constant of

$$g_{\pi NN} = 11.5. \tag{34}$$

This is closer to the experimental value

$$g_{\pi NN}|_{exp} = 12.5, \tag{35}$$

than we have any right to expect.

A similar procedure can be done for the ρ-nucleon coupling constant, with the result

$$\frac{g_{\rho NN}}{g_{\pi NN}} = 0.67, \tag{36}$$

which is not too far from the experimental value

$$\left. \frac{g_{\rho NN}}{g_{\pi NN}} \right|_{exp} = 0.46. \tag{37}$$

Likewise for the $\Delta N\pi$ coupling

$$\frac{g_{\Delta N\pi}}{g_{NN\pi}} = 1.7, \tag{38}$$

which is the $SU(6)$ result, because I am working in an $SU(6)$-like approximation ($m_\Delta = m_N$). Experiment has

$$\left. \frac{g_{\Delta N\pi}}{g_{NN\pi}} \right|_{exp} = 2.3. \tag{39}$$

That these results look reasonable without any arbitrary adjustment is encouraging. We really should not expect good agreement as there is considerable physics which has been left out (such as the effect of spin splitting in the wavefunctions of the π and the ρ, N and Δ, fissioning of bags, etc.). However, the results may be indicating that this is a reasonable first approximation for the intercoupling of states with different numbers of quarks.

Even if we can begin to understand coupling constants in this way,

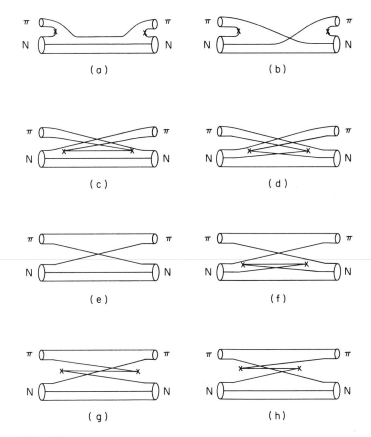

Figure 1
Some diagrams contributing to πN scattering in the quark model. The x indicates a factor of β_{ij} or ε_{ij} in the transformation of eqs. (12) or (14).

we are far from understanding pion-nucleon scattering. However, it appears that there can be differences between a quark-model approach and the Chew-Low theory. In figure 1 I have drawn the various diagrams with up to two factors of the relabeling of quarks and antiquarks (i.e., two factors of β_{ij} or ε_{ij}). I am assuming that the $\pi N \to \pi N$ transition can be expanded in factors of β_{ij} and ε_{ij} instead of (but somewhat similar to) the usual expansion in terms of a coupling constant. In the figure, factors of β or ε are denoted by an x; other lines running from initial to final states imply quark overlaps similar to α_{ij} or γ_{ij} (i.e., diagonal in $q \to q$ or $\bar{q} \to \bar{q}$). The first four digrams are similar to the usual Born amplitudes, with a, b being the direct coupling and c, d being the crossed diagrams.

In addition to coupling to the nucleon in the intermediate state, one would also have the Δ and other baryon resonances.

Diagram (e) is not present in the usual analysis. It appears to be a contribution to s-wave scattering, as it does not have any p-wave character (in contrast to, say, $a \rightarrow d$). It has *no* factors of β or ε and it would be nonvanishing even if the π bag and the nucleon bag were centered on the same point. It appears to me that the intermediate state in this case is best described as a $4q1\bar{q}$ "cryptoexotic" state. There are no couplings (i.e., β or ε) because the $4q1\bar{q}$ state can simply fall apart by rearrangement of its quarks. These have been analyzed [3] by the P-matrix techniques of Jaffe and Low [4]. Diagrams f, g, h look like corrections to diagram e, of second order in β and ε. Here care will be needed to avoid double counting and to properly define "renormalized" quantities. (For example, under certain conditions, diagram f may be identical to diagram e by a completeness relation.)

At present I have no idea how to calculate these diagrams or to relate them to πN scattering. Previous quark-model calculations have always dealt with bound states and operators all defined at one common time. For the πN scattering diagrams one has necessarily different times and different positions for the various particles and one needs to understand the propagation of the bound states. Perhaps what is needed is a way to define quark-model states with definite momentum. In a relativistic theory this is much more complicated than in nonrelativistic physics, and the appropriate prescription has not yet been given. However, complicated does not mean impossible, and there is hope that we can "catch up" to the work of Chew and Low and begin to understand πN scattering in terms of quarks.

References

[1] G. F. Chew and F. E. Low, *Phys. Rev.* 101: 1570 (1956).

[2] This presentation is based on J. F. Donoghue, *Phys. Rev. D* 25: 854 (1982).

[3] C. Roiesnel, *Phys. Rev. D* 20: 1646 (1980).

[4] R. L. Jaffe and F. E. Low, *Phys. Rev. D* 19: 2105 (1979).

The Simplicity of Bremsstrahlung Cross Sections

D. Danckaert,
P. DeCausmaecker,
R. Gastmans, W. Troost,
and
Tai Tsun Wu

Calculating helicity amplitudes is shown to be a practical method for obtaining simple cross-section formulae for bremsstrahlung processes, in particular for $e^+e^- \to 4$ jets.

Radiative processes have always played an important role in elementary-particle physics. Before the advent of the e^+e^- storage rings, most of the radiative processes studied were decay processes, such as $\mu^- \to e^- \bar{v}_e v_\mu \gamma$ or $K^0 \to \pi^+ l^- \bar{v}\gamma$ ($l = e$ or μ). Their study provided much detailed information about the structure of the weak currents and the off-shell behavior of form factors.

In e^+e^- physics at high energies, processes like $e^+e^- \to e^+e^-\gamma$, $\mu^+\mu^-\gamma$, and $\gamma\gamma\gamma$ also play an important role. The knowledge of the cross section for $e^+e^- \to e^+e^-\gamma$ (radiative Bhabha scattering) is essential for the precise determination of the beam luminosity. Radiative muon production is an important background for the charge asymmetry due to weak effects in $e^+e^- \to \mu^+\mu^-$ [1], and $e^+e^- \to \gamma\gamma\gamma$ can also act as a background for the production of hadronic channels in e^+e^- annihilation.

More recently, with the advent of quantum chromodynamics as a possible theory of the strong interactions, radiative processes in which one or more gluons are emitted have received much attention as well. The observed $e^+e^- \to 3$ jet events were interpreted, within the framework of perturbative QCD, as manifestations of the subprocess $e^+e^- \to q\bar{q}g$ [2] and led to the experimental discovery of the gluon [3].

From the theoretical point of view, radiative processes are, however, more complicated than nonradiative processes. Usually they are described in terms of many Feynman diagrams. Although the standard procedures for squaring Feynman amplitudes and for summing over the polarization degrees of freedom are well known, they often lead to very lengthy expressions which do not provide us with much insight into the structure of the cross sections. Moreover, by making use of momentum conservation, they can be expressed in many different ways, and it is not a priori clear what are the best variables to use if one wants to obtain simple formulae.

The first attempt to organize the calculation of radiative processes in a systematic way is due to Low [4]. Making use of soft-photon theorems and the gauge invariance of the theory, he was able to relate the radiative matrix element to the nonradiative matrix element and its derivative with respect to the photon momentum. Most of the applications of Low's theorem were made to radiative decay processes.

Today's high-energy physics has, however, the advantage that the energies involved in the collisions are so high that in most cases the masses of the leptons and the quarks can be neglected, at least in a first approximation. One can then expect that great simplifications should occur in the computations and that elegant formulae can be obtained for radiative collision processes.

This is, of course, easy to say. But what kind of simple formulae should one be looking for? It took several months of hard work and much struggling with long traces of γ matrices to find out that the single-bremsstrahlung cross sections for the QED processes $e^+e^- \to \mu^+\mu^-\gamma$, $e^+e^-\gamma$, and $\gamma\gamma\gamma$ had an amazingly simple structure in the massless limit [5]. They all factorized into two factors. One factor was formally the same as the well-known infrared factor but, of course, not to be evaluated in the soft-photon limit as we are dealing with hard photon emission. The second factor was nothing else but the squared matrix element for the nonradiative process expressed in the appropriate variables.

Once this result was established, it became possible to conjecture the cross sections for QCD processes like $qq \to qqg$, $q\bar{q} \to ggg$, and $gg \to ggg$ along these lines and to verify whether they agreed with the formulae obtained through the standard procedures of covariant spin summations, etc.

Simple formulae should, however, be obtained in a simple way. This is where the helicity-amplitude formalism comes in. Indeed, if it is possible to obtain simple expressions for the helicity amplitudes, it becomes a trivial matter to obtain a simple formula for the cross section as well. A key step in this direction was the introduction, in a covariant way, of explicit polarization vectors for the radiated gauge particles [6].

Consider, for example, a photon with four-momentum k which is radiated from a charged fermion line for which q_- and q_+ are the momenta of the outgoing fermion and antifermion. We can then construct two photon polarizations orthogonal to k:

$$\varepsilon_\mu^\parallel = N[(q_+ k)q_{-\mu} - (q_- k)q_{+\mu}], \quad \varepsilon_\mu^\perp = N\varepsilon_{\mu\alpha\beta\gamma}q_+^\alpha q_-^\beta k^\gamma, \tag{1}$$

where the normalization factor is

$$N = [2(q_+ q_-)(q_+ k)(q_- k)]^{-1/2}. \tag{2}$$

More interesting is the introduction of circularly polarized states

$$\varepsilon_\mu^\pm = \frac{1}{\sqrt{2}}(\varepsilon_\mu^\parallel \pm i\varepsilon_\mu^\perp). \tag{3}$$

This is due to the fact that, in QED, only the combination $\not{\varepsilon}^\pm$ appears, which can be written in the form

$$\not{\varepsilon}^\pm = -\frac{1}{2\sqrt{2}}N[\not{k}\not{q}_-\not{q}_+(1 \pm \gamma_5) - \not{q}_-\not{q}_+\not{k}(1 \mp \gamma_5)], \tag{4}$$

where we have dropped terms of the type $(q_+ q_-)\not{k}$, which vanish because of current conservation.

From this expression it is immediately clear that if the helicities of the massless fermions are also fixed, only one of the terms can contribute. Furthermore, if $\not{\varepsilon}^\pm$ stands next to a spinor $\bar{u}(q_-)$ or $v(q_+)$, this one term will either give zero because of the Dirac equations $\bar{u}(q_-)\not{q}_- = 0$ or $\not{q}_+ v(q_+) = 0$, or give rise to a factor $2(q_\pm \cdot k)$ that cancels the denominator of the adjacent fermion propagator.

In this way denominators get canceled that reappear in the normalization factor N, but as an overall factor. Most of the previous work involved in obtaining simple single-bremsstrahlung cross sections consisted of writing the different contributions over a common denominator. It is seen that the above introduction of the photon polarization vectors in the helicity amplitudes circumvents this recombination in an elegant way.

With some minor modifications, this technique can also be applied advantageously to the case of gluon bremsstrahlung. The net result is always a simple expression for the various helicity amplitudes for single bremsstrahlung, which exhibits moreover the factorization property mentioned before. One readily realizes that it is much easier to find this factorization at the level of amplitudes than it is for the cross section!

Spurred by these encouraging results, it was almost compelling to verify whether similar simplifications would occur in multiple-bremsstrahlung processes as well. In the examination of double-bremsstrahlung helicity amplitudes, it appeared, however, that for some of them the cancellation of the denominators of the fermion propagators was only partial. Propagators of the type $(q + k + l)^{-2}$, where q is an external fermion momentum and k and l photon momenta, did not always cancel, and writing everything over a common denominator did not lead to simplifications.

Once it became clear that double-bremsstrahlung processes could not be described in terms of simple cross-section formulae, the best thing then to do was to get the simplest expressions possible for the helicity amplitudes. This was accomplished in two steps. First, one eliminates the repeated indices, i.e., the $\ldots \gamma_\mu \ldots \gamma^\mu \ldots$-type expressions, where one γ_μ is associated with a fermion line, and the other γ^μ with another fermion line, both being connected by a virtual photon (or gluon) propagator. Secondly, one introduces the explicit expressions for the spinors in terms of the momenta and multiplies out the spinors and γ matrices. In this way, the helicity amplitudes are ultimately written as complex functions in the components of the various momenta in the process.

Let us clarify this with a simple example [7]. Consider first the following expression, which contains repeated indices:

$$M = \bar{u}(k_1)\gamma_\mu(1 - \gamma_5)u(k_2)\bar{u}(k_3)\gamma^\mu(1 + \gamma_5)u(k_4), \tag{5}$$

where k_i ($i = 1, 2, 3, 4$) are lightlike four-vectors. Obviously,

$$M = \frac{\bar{u}(k_1)\gamma_\mu(1 - \gamma_5)u(k_2)\bar{u}(k_2)(1 + \gamma_5)u(k_3)\bar{u}(k_3)\gamma^\mu(1 + \gamma_5)u(k_4)}{\bar{u}(k_2)(1 + \gamma_5)u(k_3)}. \tag{6}$$

But, as we are dealing with massless particles, we have

$$(1 - \gamma_5)u(k_2)\bar{u}(k_2)(1 + \gamma_5) = (1 - \gamma_5)\sum_h u(k_2)\bar{u}(k_2)(1 + \gamma_5)$$
$$= 2\not{k}_2(1 + \gamma_5), \tag{7}$$

where the summation runs over both helicities associated with a spinor $u(k_2)$. In the same way, we can effectively replace $u(k_3)\bar{u}(k_3)$ by $\not{k}_3$ in this expression. Hence,

$$\begin{aligned}
M &= 4\frac{\bar{u}(k_1)\gamma_\mu\not{k}_2\not{k}_3\gamma^\mu(1 + \gamma_5)u(k_4)}{\bar{u}(k_2)(1 + \gamma_5)u(k_3)} \\
&= 16(k_2k_3)\frac{\bar{u}(k_1)(1 + \gamma_5)u(k_4)}{\bar{u}(k_2)(1 + \gamma_5)u(k_3)} \\
&= 16(k_2k_3)\frac{\bar{u}(k_1)(1 + \gamma_5)u(k_4)\bar{u}(k_3)(1 - \gamma_5)u(k_2)}{\mathrm{Tr}\left[\not{k}_2(1 + \gamma_5)\not{k}_3(1 - \gamma_5)\right]} \\
&= 2\bar{u}(k_1)(1 + \gamma_5)u(k_4)\bar{u}(k_3)(1 - \gamma_5)u(k_2).
\end{aligned} \tag{8}$$

This expression no longer contains repeated indices.

In the second step, we introduce explicit spinors [8]. Suppose that we fix our coordinate axes, and that we choose a representation of the γ matrices for which

$$\gamma_5 = \begin{pmatrix} 1 & 0 \\ 0 & -1 \end{pmatrix}. \tag{9}$$

Then we can take, for any lightlike vector k,

$$u_+(k) = v_-(k) = \begin{pmatrix} \sqrt{k_+} \\ \sqrt{k_-}e^{i\phi_k} \\ 0 \\ 0 \end{pmatrix}, \quad u_-(k) = v_+(k) = \begin{pmatrix} 0 \\ 0 \\ -\sqrt{k_-}e^{-i\phi_k} \\ \sqrt{k_+} \end{pmatrix}, \tag{10}$$

where we have introduced the notation

$$k_\pm = k_0 \pm k_z, \quad k_\perp = k_x + ik_y = |k_\perp|e^{i\phi_k}. \tag{11}$$

Clearly, the first spinor is an eigenstate of $1 + \gamma_5$, and the second one of $1 - \gamma_5$.

With these formulae, it is easy to see that

$$\bar{u}(k_i)(1 + \gamma_5)u(k_j) = [\bar{u}(k_j)(1 - \gamma_5)u(k_i)]^* = 2\frac{k_{j\perp}Z_{ij}^*}{k_{j-}\sqrt{k_{i+}k_{j+}}}, \tag{12}$$

where

$$Z_{ij} = k_{i+}k_{j-} - k_{i\perp}^*k_{j\perp}, \quad i = 1, 2, 3, 4. \tag{13}$$

By virtue of the relation

$$2\not{k}_i(1 \pm \gamma_5) = (1 \mp \gamma_5)u(k_i)\bar{u}(k_i)(1 \pm \gamma_5), \tag{14}$$

we can reduce any spinor expression to a product of simple expressions of the foregoing type. Thus, after the elimination of the repeated indices, we can express any helicity amplitude in terms of the components of the four-vectors in the process.

This program was applied to the case of 4-jet production in e^+e^- annihilation [5]. In the framework of perturbative QCD, one has to evaluate the cross sections for $e^+e^- \to q\bar{q}q\bar{q}$ and $e^+e^- \to q\bar{q}gg$, which is a double-bremsstrahlung process.

For the first subprocess, it was found that all nonzero helicity amplitudes could be expressed in terms of a single complex function,

$$F(1, 2, 3, 4) = \frac{k_{4\perp}}{(k_3 k_4)k_{4-}} \left[\frac{k_{2+}}{E - k_{20}} Z_{14}^* (Z_{31} + Z_{34}) \right.$$
$$\left. + \frac{k_{1\perp} k_{3\perp}^*}{(E - k_{10})k_{3-}} Z_{23}(Z_{24}^* + Z_{34}^*) \right], \tag{15}$$

where the four-vectors k_i ($i = 1, 2, 3, 4$) are the momenta of the outgoing quarks and E is the beam energy. We also took the positive z axis to be along $\mathbf{p}_+$, the momentum of the incoming positron.

For the case of $e^+ e^- \to q\bar{q}gg$, we found that the helicity amplitudes were of two types. A first set, for which the gluon helicities were equal, was quite simple and did not contain denominators of the type $E - k_{10}$. The second set, however, was more complicated and did contain these denominators.

To obtain the 4-jet cross section, one must sum all the squared absolute values of the helicity amplitudes, perform the color sum, average over the initial lepton helicities, symmetrize appropriately in the four-momenta of the outgoing particles, and sum over the quark flavors.

For the subprocess $e^+ e^- \to q\bar{q}q\bar{q}$, we find

$$\overline{|M_q|^2} = \frac{N^2 - 1}{4N} A \left(\sum_f Q_f^2 \right) \sum_p [2NN_f F(1, 2, 3, 4) + F(1, 3, 2, 4)$$
$$+ F(4, 2, 3, 1)]F^*(1, 2, 3, 4), \tag{16}$$

where

$$A = e^4 g^4 [16E^4 k_{1+} k_{2+} k_{3+} k_{4+}]^{-1}, \tag{17}$$

and e is the electron charge, g the $SU(N)$ gauge coupling constant, Q_f the fractional quark charges, and N_f the number of quark flavors. The first summation sums over all quark flavors, while the second summation sums over all 24 permutations of the jet momenta k_1, k_2, k_3, and k_4.

Similarly, for $e^+ e^- \to q\bar{q}gg$, we find

$$\overline{|M_g|^2} = \frac{N^2 - 1}{4N} B \left(\sum_f Q_f^2 \right) \sum_p G(1, 2, 3, 4), \tag{18}$$

where

$$B = e^4 g^4 [(k_1 k_2)(k_1 k_3)(k_1 k_4)(k_2 k_3)(k_2 k_4)(k_3 k_4)]^{-1}, \tag{19}$$

$$G(1, 2, 3, 4) = (k_3 k_4) \left\{ (k_{4+}^2 + k_{4-}^2)[2N^2(k_1 k_3)(k_2 k_4) - (k_1 k_2)(k_3 k_4)] \right.$$
$$\left. + \frac{(N^2 - 2)|c_1 + c_2|^2 + N^2 |c_1 - c_2|^2}{32E^2 (k_1 k_2)k_{1+}^2 k_{2-}^2 k_{3+} k_{4-}} \right\}, \tag{20}$$

with

$$c_1 = |\alpha|^2 + \frac{(k_2 k_4)}{2E(E - k_{30})}\alpha\beta + \frac{(k_1 k_3)}{2E(E - k_{40})}\alpha^*\gamma,$$

$$c_2 = \beta\gamma + \frac{(k_1 k_4)}{2E(E - k_{30})}\alpha\beta + \frac{(k_2 k_4)}{2E(E - k_{40})}\alpha^*\gamma,$$

$$\alpha = Z_{12}Z_{34} - Z_{14}Z_{32},$$

$$\beta = Z_{32}^*[Z_{12}^* + Z_{14}^*],$$

$$\gamma = Z_{14}[Z_{12} + Z_{32}].$$

(21)

Including phase space factors, we thus obtain for the 4-jet cross section,

$$d\sigma(4\text{-jet}) = \frac{\delta^4(p_+ + p_- - k_1 - k_2 - k_3 - k_4)}{128E^2(2\pi)^8}[\overline{|M_q|^2} + \overline{|M_g|^2}]$$

$$\frac{d^3\mathbf{k}_1 d^3\mathbf{k}_2 d^3\mathbf{k}_3 d^3\mathbf{k}_4}{k_{10}k_{20}k_{30}k_{40}},$$

(22)

with p_+ (p_-) being the four-momentum of the incoming positron (electron).

Because of the relative simplicity of our formulae, we are able to establish that the 4-jet cross section is proportional to $R = \sum_f Q_f^2$. The terms proportional to $(\sum_f Q_f)^2$ dropped out after momentum symmetrization.

It is our hope that the above procedures can be applied systematically to all multiple-bremsstrahlung processes in gauge theories at high energies. In view of the relative strength of the strong coupling constant, one could indeed imagine that even higher-order processes like $e^+ e^- \to 5$ jets have to be calculated. In that case, it is very unlikely that the standard procedures of covariant summation of Feynman amplitudes could be carried out, even with the help of algebraic manipulation programs on the computer. There are simply too many Feynman diagrams. On the other hand, we see no reason why the procedure outlined in this paper would fail.

Acknowledgments

This work was supported in part by NATO research grant RG 079.80 and by U.S. Department of Energy contract DE-ASO2-76ERO3227.

References

[1] JADE Collaboration, W. Bartel et al., DESY Preprint 81-072 (to be published); TASSO Collaboration, R. Brandelik et al., DESY Preprint 82-002 (to be published).

[2] J. Ellis, M. K. Gaillard, and G. G. Ross, *Nucl. Phys. B* 111: 253 (1976); T. A. DeGrand, Y. J. Ng, and S.-H. Tye, *Phys. Rev. D* 16: 3251 (1977); A. DeRujula, J. Ellis, E. G. Floratos, and M. K. Gaillard, *Nucl. Phys. B* 138: 387 (1978).

[3] B. H. Wiik, Proc. Int. Neutrino Conf. (Bergen, Norway, June 1979), p. 113; P. Söding, Proc. EPS Intern. Conf. on High Energy Physics (CERN, June 1979), p. 271; TASSO Collaboration, R. Brandelik et al., *Phys. Lett. B* 86: 243 (1979); MARK-J Collaboration, D. P. Barber et al., *Phys. Rev. Lett.* 43: 830 (1979); PLUTO Collaboration, Ch. Berger et al., *Phys. Lett. B* 86: 418 (1979); JADE Collaboration, W. Bartel et al., *Phys. Lett. B* 91: 142 (1980).

[4] F. E. Low, *Phys. Rev.* 110: 974 (1958).

[5] F. A. Berends, R. Kleiss, P. DeCausmaecker, R. Gastmans, and T. T. Wu, *Phys. Lett. B* 103: 124 (1981).

[6] P. DeCausmaecker, R. Gastmans, W. Troost, and T. T. Wu, *Phys. Lett. B* 105: 215 (1981), and DESY Preprint DESY 81-050 (1981).

[7] F. A. Berends, R. Kleiss, P. DeCausmaecker, R. Gastmans, W. Troost, and T. T. Wu, Univ. Leuven Preprint KUL-TF-81/18 (1981).

[8] D. Danckaert, P. DeCausmaecker, R. Gastmans, W. Troost, and T. T. Wu, Univ. Leuven Preprint KUL-TF-82/1 (1982).

Gauge Invariance and Mass, III

R. Jackiw

Gauge invariance of a vector field need not imply that the vector particle is massless. Two completely solvable models that exemplify this are reviewed: the two-dimensional Schwinger model, and three-dimensional massive electrodynamics. In contrast to previous discussions, topological mechanisms for mass generation are emphasized. In two dimensions the Pontryagin density, and in three, the Chern-Simons secondary characteristic class are identified as the relevant structures responsible for the appearance of a gauge-invariant mass. The latter, in a non-Abelian theory, provides a quantization condition on the mass. Quantum field–theoretic perturbation theory for the three-dimensional model is both ultraviolet and infrared finite, without renormalization.

I chose a serial title for this essay to highlight its connection to two papers by Julian Schwinger, named similarly but numbered I and II, in which he showed that massive gauge fields need not violate gauge invariance [1]. He exemplified the general ideas in a two-dimensional model—paper II of the series—while I shall be concerned with a model in three dimensions; thus, another reason for "III."

First, let us review the two-dimensional Schwinger model, not only to set the stage for our three-dimensional discussion, but also to demonstrate that the emergence of mass can be seen from very general arguments, which are established without solving the theory.

The Schwinger model concerns quantum electrodynamics in two dimensions, where the Abelian gauge field couples to massless fermions:

$$\partial_\mu F^{\mu\nu} = ej^\nu,$$
$$F^{\mu\nu} = \partial^\mu A^\nu - \partial^\nu A^\mu. \tag{1}$$

The field tensor's antisymmetry requires, for consistency of (1), current conservation:

$$j^\nu = -\bar{\psi}\gamma^\nu\psi,$$
$$\partial_\nu j^\nu = 0. \tag{2}$$

Let us rewrite (1) by using the dual tensor, which in two dimensions is a scalar:

$$*F = \frac{1}{2}\varepsilon^{\mu\nu}F_{\mu\nu},$$
$$F^{\mu\nu} = -\varepsilon^{\mu\nu}*F. \tag{3}$$

In terms of $*F$, eq. (1) reads

$$\partial_\mu \varepsilon^{\mu\nu}*F = -ej^\nu. \tag{4}$$

The epsilon tensor may be removed by introducing the dual current:

$$*j^\mu = \varepsilon^{\mu\nu}j_\nu, \tag{5}$$
$$\partial_\mu *F = e*j_\mu. \tag{6}$$

Upon taking another divergence of (6), we obtain

$$\Box *F = e\partial_\mu *j^\mu. \tag{7}$$

We recognize that a mass term, if any, for the field $*F$ is governed by the divergence of the dual current.

In two dimensions, we can go further and determine properties of the dual current. It is a feature of two-dimensional Dirac algebra that the axial vector is dual to the vector:

$$\varepsilon^{\mu\nu}\gamma_\nu = i\gamma^\mu\gamma^5,$$
$$\gamma^5 = i\gamma^0\gamma^1. \tag{8a}$$

Hence the dual current is just the axial vector current,

$$*j^\mu = j_5^\mu = \bar\psi i\gamma^5\gamma^\mu\psi, \tag{8b}$$

and with massless fermions one expects it to be conserved, owing to chiral invariance.

However, just as in four dimensions [2], there is an anomaly [3]. While the anomaly is usually presented as a consequence of regularization procedures needed to remove perturbation-theory infinities, in the present context we can recognize its necessary occurrence without any graphical computation, since a nontrivial two-dimensional vector operator which is conserved and also satisfies $\partial_\mu \varepsilon^{\mu\nu}j_\nu = 0$ cannot exist. More explicitly, let us consider a two-vector current correlation function

$$\langle 0|j^\mu(x)j^\nu(y)|0\rangle = \left(g^{\mu\nu} - \frac{\partial^\mu\partial^\nu}{\Box}\right)\Pi(x-y). \tag{9}$$

The transverse form is dictated by current conservation (2). But two-dimensional geometry relates the axial vector current–vector current correlation function to (9):

$$\langle 0|j_5^\mu(x)j^\nu(y)|0\rangle = \left(\varepsilon^{\mu\nu} - \varepsilon^{\mu\alpha}\frac{\partial_\alpha\partial^\nu}{\Box}\right)\Pi(x-y). \tag{10}$$

It is now clear that j_5^μ cannot be conserved:

$$\langle 0|\partial_\mu j_5^\mu(x)j^\nu(y)|0\rangle = -\varepsilon^{\nu\mu}\partial_\mu\Pi(x-y) \neq 0. \tag{11}$$

What is the form of the anomalous divergence? Equation (11) puts into evidence the coupling of $\partial_\mu j_5^\mu$ to a one-photon state, in contrast to four dimensions, where such coupling is prohibited by charge conjugation, and occurs only on the two-photon level. Consequently, $\partial_\mu j_5^\mu$ will involve one power of e, and the only operator with proper dimensionality and quantum numbers is $*F$. Hence $\partial_\mu j_5^\mu$ must be proportional to $e*F$, and an explicit calculation is needed only to determine the numerical proportionality constant, necessarily negative owing to the positivity conditions satisfied by (9). This computation gives the well-known result [3]

$$\partial_\mu j_5^\mu = -\frac{e}{\pi}*F, \tag{12}$$

which when combined with (7), shows the "photon" to be massive [4]:

$$\left(\Box + \frac{e^2}{\pi}\right)*F = 0. \tag{13}$$

Note that (7) may be written as

$$\Box F_{\mu\nu} = e(\partial_\mu j_\nu - \partial_\nu j_\mu). \tag{14}$$

This formula demonstrates that a mass is generated if the vector current satisfies a London ansatz:

$$j_\mu = -\frac{e}{\pi}A_\mu + \cdots . \tag{15}$$

[The dots indicate other possible operators.] While (15) is consistent with (12), which is equivalent to

$$\partial_\mu j_\nu - \partial_\nu j_\mu = -\frac{e}{\pi}F_{\mu\nu}, \tag{16}$$

we see that the present development does not use the London ansatz. Rather, (16) [or (12)] follows directly from the axial vector current anomaly, which these days we understand to be a consequence of topological properties of gauge fields (see below).

Let us now turn to three dimensions, where it is possible to construct a local gauge field Lagrangian that leads to a gauge-invariant mass for the gauge fields [5]. The Lagrangian for an Abelian theory is

$$\mathcal{L} = -\frac{1}{4}F^{\mu\nu}F_{\mu\nu} + \frac{\mu}{4}\varepsilon^{\mu\nu\alpha}F_{\mu\nu}A_{\alpha},$$
$$F_{\mu\nu} = \partial_{\mu}A_{\nu} - \partial_{\nu}A_{\mu}.$$

(17)

The field equations may be presented in "Maxwell" form, without reference to the vector potential, by introducing the dual tensor, which here is a vector:

$$\partial_{\mu}F^{\mu\nu} + \mu * F^{\nu} = 0,$$

(18)

$$*F^{\mu} = \frac{1}{2}\varepsilon^{\mu\alpha\beta}F_{\alpha\beta}, \quad F^{\mu\nu} = \varepsilon^{\mu\nu\alpha}*F_{\alpha}.$$

(19)

The (Bianchi) identity

$$\partial_{\mu}*F^{\mu} = 0,$$

(20)

is not independent, but follows from (18). Clearly, the field equations are gauge invariant. The Lagrangian is not; under a gauge transformation,

$$A_{\mu} \rightarrow A_{\mu} + \partial_{\mu}\theta,$$

(21)

it changes by a total derivative:

$$\mathcal{L} \rightarrow \mathcal{L} + \frac{\mu}{2}\partial_{\alpha}(*F^{\alpha}\theta).$$

(22)

But the action is invariant:

$$\int dx\mathcal{L} \rightarrow \int dx\mathcal{L} + \frac{\mu}{2}\int dx\partial_{\alpha}(*F^{\alpha}\theta) = \int dx\mathcal{L},$$

(23)

since the field strengths fall off at large distances, and we assume that θ attains a unique limit at $t = \pm\infty$.

To show that μ is indeed a mass, we proceed through the same steps as in the Schwinger model. Equation (18) is written in terms of dual variables, and the mass term is taken to define a current J^{ν}:

$$\partial_{\mu}\varepsilon^{\mu\nu\alpha}*F_{\alpha} = \mu J^{\nu},$$

(24)

$$J^{\nu} \equiv -*F^{\nu}.$$

(25)

The epsilon tensor is removed and a dual current is introduced:

$$\partial_\mu {}^*F_\nu - \partial_\nu {}^*F_\mu = -\mu {}^*J_{\mu\nu}, \tag{26}$$

$${}^*J_{\mu\nu} = \varepsilon_{\mu\nu\alpha}J^\alpha. \tag{27}$$

Finally, another divergence of (26) shows, with the help of (20), that a possible mass term is determined by the divergence of the dual current:

$$\square {}^*F^\nu = -\mu \partial_\mu {}^*J^{\mu\nu}. \tag{28}$$

In this model, the dual current is just the field strength itself. From definitions (25) and (27), we have according to (19)

$${}^*J_{\mu\nu} = -\varepsilon_{\mu\nu\alpha} {}^*F^\alpha. \tag{29}$$

Hence, the field equation (18) gives the divergence of ${}^*J_{\mu\nu}$:

$$\partial_\mu {}^*J^{\mu\nu} = -\partial_\mu F^{\mu\nu} = \mu {}^*F^\nu. \tag{30}$$

Consequently, (28) shows that the gauge field is massive:

$$(\square + \mu^2) {}^*F^\nu = 0. \tag{31}$$

Equation (28) may also be presented as

$$\square F_{\mu\nu} = \mu(\partial_\mu J_\nu - \partial_\nu J_\mu); \tag{32}$$

and (25), (26), and (29) imply

$$\partial_\mu J_\nu - \partial_\nu J_\mu = -\mu F_{\mu\nu}. \tag{33}$$

But again, we see that a London ansatz is not initially responsible for (33). Rather, a topological effect is at work, which we shall explain below, when massive non-Abelian gauge fields are discussed. Consequently, we shall call our mass term "topological."

The physical content of the theory is further exhibited after quantization [6]. Canonical commutators, determined (in any gauge) from (17), lead to the following gauge-invariant equal-time algebra:

$$\begin{aligned} i[E^i(\mathbf{x}), E^j(\mathbf{y})] &= \mu\varepsilon^{ij}\delta(\mathbf{x} - \mathbf{y}), \\ i[E^i(\mathbf{x}), B(\mathbf{y})] &= -\varepsilon^{ij}\partial_j\delta(\mathbf{x} - \mathbf{y}), \\ i[B(\mathbf{x}), B(\mathbf{y})] &= 0. \end{aligned} \tag{34}$$

The common time is suppressed in the arguments of the electromagnetic fields, given by the usual components of the field tensor:

$$F^{i0} = E^i, \quad F^{ij} = -\varepsilon^{ij}B. \tag{35}$$

The commutation relations (34), as well as the field equations (18)–(20) are solved by a canonical, free, massive field ϕ,

$$(\Box + \mu^2)\phi = 0,$$
$$i[\dot{\phi}(\mathbf{x}), \phi(\mathbf{y})] = \delta(\mathbf{x} - \mathbf{y}),$$
(36)

in terms of which the electromagnetic fields are simply expressed:

$$E^i = \varepsilon^{ij}\hat{\partial}_j\dot{\phi} + \mu\hat{\partial}_i\phi,$$
$$B = -\sqrt{-\nabla^2}\,\phi,$$
(37)
$$\hat{\partial}_i \equiv \frac{\partial_i}{\sqrt{-\nabla^2}}.$$

The energy-momentum tensor is conventional:

$$\theta^{\mu\nu} = -F^{\mu\alpha}F^\nu{}_\alpha + \frac{g^{\mu\nu}}{4}F^{\alpha\beta}F_{\alpha\beta}.$$
(38)

The topological mass does not appear here, since in the presence of an external gravitational field $g^{\mu\nu}$, the expression $(\mu/4)\int dx\varepsilon^{\mu\nu\alpha}F_{\mu\nu}A_\alpha$ is generally covariant without any additional metric factors. Hence a variation with respect to $g^{\mu\nu}$, which determines $\theta_{\mu\nu}$, does not see the mass term. (This is one aspect of the topological features of the theory.) Therefore, the Hamiltonian takes the usual electromagnetic expression,

$$H = \tfrac{1}{2}\int d\mathbf{x}(\mathbf{E}^2 + B^2);$$
(39)

in terms of ϕ the Hamiltonian is

$$H = \tfrac{1}{2}\int d\mathbf{x}(\dot{\phi}^2 + \phi(-\nabla^2 + \mu^2)\phi).$$
(40)

Thus we see that the theory describes a massive free particle.

Upon examining the Lorentz generators [6], one learns that the spin eigenvalue is $\mu/|\mu| = \pm 1$. (In two spatial dimensions "spin" is a rotational pseudoscalar—it can be viewed as the projection of a three-dimensional spin along the (missing) third axis.) A theory with only one particle state carrying nonvanishing spin violates parity P and time inversion T. That the topological mass does not possess these discrete symmetries is of course already apparent from the epsilon tensor in the Lagrangian. (One may consider two gauge fields, one with mass $+|\mu|$, the other with $-|\mu|$. Both spins ± 1 now occur and P and T are conserved, since coordinate inversion supplemented by field exchange leaves the action invariant.)

The massless limit is not smooth for the spin: the massless theory, which is P and T invariant, possesses a single degree of freedom with spin 0. This is consistent with the result that massless representations of the Poincaré group in three space-time dimensions do not carry a spin quantum number [7].

Although the topological mass does not arise through a Higgs mechanism, there is an interesting analogy. Consider the time component of the field equation (Gauss's law) in the presence of an external charge density ρ:

$$\partial_i E^i - \mu B = \rho. \tag{41a}$$

Upon integrating this over all space, the first term vanishes, since the fields, being massive, decrease exponentially at large distances. One is left with

$$-\mu \int d\mathbf{x}\, B = \int d\mathbf{x}\, \rho = Q. \tag{41b}$$

The flux passing out of our two-dimensional space is proportional to the external charge. Correspondingly, the magnetic potential is long-range, even though the magnetic field is short-range:

$$\mathbf{A} \xrightarrow[r\to\infty]{} -\mathbf{\nabla} \frac{Q}{2\pi\mu} \tan^{-1} \frac{y}{x}. \tag{42}$$

This is similar to the electromagnetic configuration supported by vortices in the Higgs model [8], and again highlights the topological features of our theory.

Further interesting structure of a three-dimensional gauge theory with topological mass emerges when non-Abelian fields are considered. The Lagrangian is

$$\mathcal{L} = \frac{1}{2g^2} \operatorname{Tr} F^{\mu\nu} F_{\mu\nu} - \frac{\mu}{2g^2} \varepsilon^{\mu\nu\alpha} \operatorname{Tr} \left(F_{\mu\nu} A_\alpha - \frac{2}{3} A_\mu A_\nu A_\alpha \right). \tag{43}$$

We are using a matrix notation,

$$\begin{aligned} A_\mu &= g T^a A_\nu^a, \\ F_{\mu\nu} &= g T^a F_{\mu\nu}^a = \partial_\mu A_\nu - \partial_\nu A_\mu + [A_\mu, A_\nu], \end{aligned} \tag{44}$$

which employs the representation matrices of the group:

$$[T^a, T^b] = f^{abc} T^c. \tag{45}$$

The coupling constant is g. The field equations are gauge covariant:

$$\mathcal{D}_\mu F^{\mu\nu} + \mu *F^\nu = 0,$$
$$\mathcal{D}_\mu = \partial_\mu + [A_\mu, \ ]. \tag{46}$$

The dual field

$$*F^\mu = \tfrac{1}{2}\varepsilon^{\mu\alpha\beta} F_{\alpha\beta},$$
$$F^{\mu\nu} = \varepsilon^{\mu\nu\alpha} *F_\alpha, \tag{47}$$

satisfies the (Bianchi) identity as a consequence of (44) or (46):

$$\mathcal{D}_\mu *F^\mu = 0. \tag{48}$$

The dual of (46) is

$$\mathcal{D}_\alpha *F_\beta - \mathcal{D}_\beta *F_\alpha - \mu F_{\alpha\beta} = 0, \tag{49}$$

and another covariant divergence converts this, with the help of (46) and the (Ricci) identity $[\mathcal{D}_\alpha, \mathcal{D}_\beta] = F_{\alpha\beta}$, to

$$(\mathcal{D}_\alpha \mathcal{D}^\alpha + \mu^2) *F_\mu = \varepsilon_{\mu\alpha\beta}[*F^\alpha, *F^\beta]. \tag{50}$$

Equations (43)–(50) are the non-Abelian analogues of (17)–(31).

While we cannot solve this nonlinear theory classically or quantum mechanically, consideration of the noninteracting ($g = 0$) limit indicates that we are again dealing with massive gauge fields. However, gauge invariance puts further constraints on the mass term.

The field Lagrangian (43) is not invariant against gauge transformations; rather, just as in the Abelian theory, it changes by a total derivative. Consider a finite transformation

$$A_\mu \rightarrow U^{-1} A_\mu U + U^{-1} \partial_\mu U. \tag{51}$$

The response of the action to gauge transformations is

$$\int dx \mathcal{L} \rightarrow \int dx \mathcal{L} + \frac{\mu}{g^2} \int dx \varepsilon^{\alpha\mu\nu} \, \mathrm{Tr} \, \partial_\mu [A_\alpha \partial_\nu U U^{-1}]$$
$$+ \frac{\mu}{3g^2} \int dx \varepsilon^{\alpha\beta\gamma} \, \mathrm{Tr} \, [\partial_\alpha U U^{-1} \partial_\beta U U^{-1} \partial_\gamma U U^{-1}]. \tag{52}$$

We shall only consider gauge transformations which tend to the identity at temporal and spatial infinity:

$$U(x) \xrightarrow[x\to\infty]{} I. \tag{53}$$

This restriction is made plausible by the assumption that the ultimate reaches of space and time should be without structure. Also, the last integral in (52) may not converge, unless (53) holds.

With (53), we may conclude that the second term on the right-hand side in (52)—the analogue of (22) occurring in the Abelian theory—vanishes. The last term also can be converted to a surface integral once the integrand is rewritten as a total derivative. This is achieved by introducing an explicit parametrization for U. We choose the gauge group to be $SU(2)$ (more generally, we may consider a $SU(2)$ subgroup of the gauge group), and use the exponential parametrization:

$$U(x) = \exp i\sigma^a \theta^a(x),$$
$$\sigma^a = \text{Pauli matrices.} \tag{54}$$

It follows that

$$\int dx \mathscr{L} \rightarrow \int dx \mathscr{L} + \mu \frac{8\pi^2}{g^2} w(U),$$

$$w(U) = \frac{1}{24\pi^2} \int dx \varepsilon^{\alpha\beta\gamma} \text{Tr} \left(\partial_\alpha U U^{-1} \partial_\beta U U^{-1} \partial_\gamma U U^{-1} \right) \tag{55}$$

$$= \frac{1}{8\pi^2} \int dx \varepsilon^{\alpha\beta\gamma} \varepsilon^{abc} \partial_\alpha \left(\theta^a \partial_\beta \theta^b \partial_\gamma \theta^c \frac{1}{\theta^2} \left[1 - \frac{\sin 2\theta}{2\theta} \right] \right).$$

It is recognized that $w(U)$ is the winding number of the gauge transformation U [9]. Consequently, the surface integral is not zero, but takes an integer value which characterizes the homotopic equivalence class to which U belongs. Only for homotopically trivial U's—those continuously deformable to the unit matrix—does $w(U)$ vanish. These considerations are familiar from the analysis of topological structure in four-dimensional Yang-Mills theory [9]. That they should reappear in the three-dimensional theory is not surprising, in view of the further mathematical/topological connections which will be drawn later.

That $w(U)$ is an integer can be established, when the integration in (55) is over Euclidean three-space, by recognizing that gauge transformations satisfying condition (53) provide a mapping of S_3 (three-dimensional space with the points at infinity identified) to S_3 (the $SU(2)$ manifold). Such mappings are characterized by integral winding numbers, whose analytic expression is given by (55), in Euclidean space. However, once we know that $w(U)$ is an integer in Euclidean space, we see that it will also be an integer in Minkowski space, since (55) is generally covariant without additional metric factors.

We conclude, therefore, that the action is not gauge invariant, but changes by $\mu(8\pi^2/g^2)w(U)$, with $w(U)$ an integer. However, it is the exponential of the action, $\exp i\int dx \mathcal{L}$, that should be gauge invariant. Otherwise the expectation of a gauge-invariant operator $\mathcal{O}$ would be undefined, as can be seen from a functional integral representation

$$\langle \mathcal{O} \rangle = Z^{-1} \int \mathcal{D}A \, \mathcal{O}(A) \exp iI(A),$$

with gauge-invariant measure $\mathcal{D}A$ and normalization factor Z. Changing variables, $A \to A^U$, where A^U is a gauge transform of A, implies that

$$\langle \mathcal{O} \rangle = \exp \left[i\mu \frac{8\pi^2}{g^2} w(U) \right] \langle \mathcal{O} \rangle,$$

which can only be tolerated if the change in the action is an integral multiple of 2π. This gives a quantization condition on the dimensionless ratio $4\pi(\mu/g^2)$:

$$4\pi \frac{\mu}{g^2} = n, \, n = 0, \, \pm 1, \, \ldots . \tag{56}$$

A Euclidean formulation leads to the same conclusion. The functional integral requires $\exp\{-\int dx \mathcal{L}\}$ to be gauge invariant, but the mass term's contribution to the action is purely imaginary; a factor of $i = \sqrt{-1}$ appears when continuation to imaginary time (Euclidean space) is performed.

The action for the topological mass term provides a physical setting for the mathematical concept of Chern-Simons secondary characteristic classes [10]. These involve the following topological ideas. In even dimensions, one may construct from gauge fields a Pontryagin density $\mathcal{P}_{2n}$, whose integral over the even-dimensional space is a topological invariant. Examples in two and four dimensions are

$$\mathcal{P}_2 = \frac{1}{2\pi} {}^*F \qquad \text{(two dimensions)}, \tag{57a}$$

$$\mathcal{P}_4 = -\frac{1}{16\pi^2} \text{Tr} \, {}^*F^{\mu\nu} F_{\mu\nu} \quad \text{(four dimensions)}. \tag{57b}$$

These gauge-invariant objects can also be written as total derivatives of gauge variant quantities:

$$\mathcal{P}_{2n} = \partial_\mu X_{2n}^\mu. \tag{58}$$

The two- and four-dimensional expressions are

$$X_2^\mu = \frac{1}{2\pi}\varepsilon^{\mu\nu}A_\nu \quad \text{(two dimensions)}, \tag{59a}$$

$$X_4^\mu = -\frac{1}{16\pi^2}\varepsilon^{\mu\alpha\beta\gamma}\,\mathrm{Tr}\,(A_\alpha F_{\beta\gamma} - \tfrac{2}{3}A_\alpha A_\beta A_\gamma) \tag{59b}$$

(four dimensions).

The Chern-Simons secondary characteristic class is gotten by integrating one component of X_{2n}^μ over the $(2n-1)$-dimensional space which does not include that component. The integral is gauge invariant against homotopically trivial gauge transformations; otherwise, it changes by the winding number of the transformation [10].

We see that both the two-dimensional and three-dimensional mass terms make use of topological structures: in two dimensions, the Pontryagin density appears as an anomalous divergence of the axial vector current; in three dimensions, the mass term is the Chern-Simons secondary characteristic.

Thus far, we have been concerned with kinematical properties of three-dimensional topologically massive, gauge fields. Let us now examine dynamics, when the gauge field is coupled to massive Fermi fields, governed by the following fermion Lagrangian.

$$\mathscr{L}_F = i\bar\psi\gamma^\mu(\partial_\mu - ieA_\mu)\psi - m\bar\psi\psi. \tag{60}$$

We shall discuss only the Abelian theory.

In three dimensions, the Dirac algebra is realized by Pauli matrices,

$$\gamma^0 = \sigma^3, \quad \gamma^1 = i\sigma^1, \quad \gamma^2 = i\sigma^2, \tag{61}$$

and ψ has two components, describing a particle (and an antiparticle) with P and T violating spin $\frac{1}{2}m/|m|$. (Massless fermions in three dimensions are spinless and conserve P and T [7].)

Evidently the topological gauge field mass and the fermion mass share the same P and T quantum numbers; they belong together and one can be generated from the other dynamically (see below).

Of course, we can study the dynamics only perturbatively; but perturbation theory in this model offers new surprises. It is based on the conventional electromagnetic vertex and electron propagator,

$$S(p) = \frac{i}{p - m}, \tag{62}$$

while the photon propagator is gauge-dependent, and contains an axial contribution arising from the topological mass. In the Coulomb gauge its form is

$$D_{\mu\nu}(p) = \frac{-i}{p^2 - \mu^2 + i\varepsilon}\left[g_{\mu\nu} + \frac{\bar{p}_\mu\bar{p}_\nu}{\mathbf{p}^2} - n_\mu n_\nu\frac{p_0^2}{\mathbf{p}^2} + i\mu\varepsilon_{\mu\nu\alpha}\frac{\bar{p}^\alpha}{\mathbf{p}^2}\right],$$

$$\bar{p}^\mu = (0, \mathbf{p}), \quad n^\mu = (1, \mathbf{0}). \tag{63a}$$

In a family of covariant gauges parametrized by a constant α, we have

$$D_{\mu\nu}(p) = \frac{-i}{p^2 - \mu^2 + i\varepsilon}\left[g_{\mu\nu} - \frac{p_\mu p_\nu}{p^2} - i\mu\varepsilon_{\mu\nu\alpha}\frac{p^\alpha}{p^2}\right] - i\alpha\frac{p_\mu p_\nu}{p^4}. \tag{63b}$$

First about divergences: there are none. The theory is superrenormalizable and only one- and two-loop (single-particle irreducible) graphs possess superficial linear and logarithmic divergences. But logarithmic divergences are eliminated by symmetric integration, while the linear divergence, which occurs only in the one-loop vacuum polarization tensor, is removed by gauge-invariant integration procedures. The mass term provides an infrared cutoff, provided no infrared divergences are created by unfortunate choice of gauge. Thus one sees that the Coulomb gauge should be avoided, owing to the $|\mathbf{p}|^{-2}$ singularity in the time-time component. However, since terms proportional to p^μ may be changed at will in a photon propagator, an infrared-safe version of (63a) is

$$D_{\mu\nu}(p) = \frac{-i}{p^2 - \mu^2 + i\varepsilon}\left[g_{\mu\nu} + i\mu\varepsilon_{\mu\nu\alpha}\frac{\bar{p}^\alpha}{\mathbf{p}^2}\right] \tag{64a}$$

Similarly, for the covariant propagators, only the Landau gauge ($\alpha = 0$) is infrared-safe:

$$D_{\mu\nu}(p) = \frac{-i}{p^2 - \mu^2 + i\varepsilon}\left[g_{\mu\nu} - \frac{p_\mu p_\nu}{p^2} - i\mu\varepsilon_{\mu\nu\alpha}\frac{p^\alpha}{p^2}\right]. \tag{64b}$$

(One could also drop the $p_\mu p_\nu$ term.) The remaining singularities—$\mathbf{p}/\mathbf{p}^2$ in (64a) or p^α/p^2 in (64b)—are integrable in low perturbative orders, and never become enhanced in higher orders [6].

Thus perturbation theory is entirely finite, provided Lorentz and gauge invariance are enforced. Nevertheless, there remains an indeterminancy. The various available Lorentz- and gauge-invariant regularizations give different values for the propagator's radiative corrections.

For example, the gauge-invariant one-loop vacuum polarization tensor, contributing to the complete photon propagator $\mathscr{D}_{\mu\nu}$, is given by [6]

$$\mathcal{D}_{\mu\nu}^{-1} = D_{\mu\nu}^{-1} - i\Pi_{\mu\nu},$$

$$\Pi_{\mu\nu}(p) = -ie^2 \int \frac{dk}{(2\pi)^3} \mathrm{Tr}\, \gamma_\mu S(p+k)\gamma_\nu S(k) + O(e^4)$$

$$= (g_{\mu\nu}p^2 - p_\mu p_\nu)\frac{e^2}{8\pi} \int_{2|m|}^{\infty} da \frac{1 + (4m^2/a^2)}{p^2 - a^2 + i\varepsilon} \qquad (65)$$

$$+ im\varepsilon_{\mu\nu\alpha}p^\alpha \frac{e^2}{2\pi} \int_{2|m|}^{\infty} da \frac{1}{p^2 - a^2 + i\varepsilon}.$$

Note that an axial structure is induced by the fermion mass, providing a parity-violating contribution to the photon propagator, even if none is present in the bare-photon Lagrangian. In the second line of (65), we have dropped a linearly divergent term proportional to $g_{\mu\nu}$. This is justified by dimensional regularization, but the Pauli-Villars procedure instructs subtracting from (65) the same expression with $|m| \to \infty$. However, the $|m| \to \infty$ limit of the last term in (65) is nonzero and the Pauli-Villars answer differs from the above by $im\varepsilon_{\mu\nu\alpha}p^\alpha(e^2/4\pi)$:

$$\Pi_{\mu\nu}^{PV}(p) = (g_{\mu\nu}p^2 - p_\mu p_\nu)\frac{e^2}{8\pi} \int_{2|m|}^{\infty} da \frac{1 + (4m^2/a^2)}{p^2 - a^2 + i\varepsilon}$$

$$+ im\varepsilon_{\mu\nu\alpha}p^\alpha \frac{e^2}{2\pi} p^2 \int_{2|m|}^{\infty} da \frac{1/a^2}{p^2 - a^2 + i\varepsilon}. \qquad (66)$$

We prefer the Pauli-Villars result (66) to the dimensionally regulated one (65), because dimensional regularization is not particularly appropriate for topological effects: $\varepsilon_{\mu\nu\alpha}$ has no generalization to continuous dimensions. Also, the axial-structure function vanishes at $p^2 = 0$ in the Pauli-Villars procedure. As a consequence, if $\mu = 0$, the gauge field remains massless, at least to this order. However, with dimensional regularization, the structure function does not vanish, and a mass is induced by the radiative correction.

Three-dimensional gauge theories supplement two-dimensional ones by providing yet another theoretical laboratory in which amusing and provocative effects are found. But what is their relevance to the physical four-dimensional world? The answer is that three-dimensional (Euclidean) theories summarize the high-temperature behavior of four-dimensional theories. In that sense massless three-dimensional gauge models are physically interesting. However, it is not clear that a topological mass will emerge from a high-temperature dimensional reduction, although one wonders whether the P- and T-violating topological effects of four-dimensional theories, which are summarized by the Pontryagin density, can induce the Chern-Simons mass term in the effective three-dimensional model.

Gauge fields in four dimensions, as in any number, satisfy

$$\Box F_{\mu v} = e(\partial_\mu j_v - \partial_v j_\mu). \tag{67}$$

Consequently, for them to be massive, it must be that

$$e(\partial_\mu j_v - \partial_v j_\mu) = -\mu^2 F_{\mu v} + \cdots. \tag{68}$$

While the London ansatz $j_\mu \propto A_\mu$ comes immediately to mind as a way of fulfilling (68), we saw in the two- and three-dimensional examples a topological effect replacing it. In four dimensions, the London ansatz, in the guise of the Higgs mechanism

$$\begin{aligned}
j_\mu &= i\phi(\partial_\mu + ieA_\mu)\phi^* - i\phi^*(\partial_\mu - ieA_\mu)\phi \\
&\approx -2e(\phi^*\phi)A_\mu \quad \text{(for constant Higgs fields),}
\end{aligned} \tag{69}$$

is the familiar device for making a gauge field massive. Attempts to replace this by a bound-state mechanism [11] have not been successful [12], so one may speculate that perhaps topology will once again solve our physics problems.

References

[1] J. Schwinger, *Phys. Rev.* 125: 397 (1962) and 128: 2425 (1962).

[2] R. Jackiw in *Lectures on Current Algebra and Its Applications*, edited by S. Treiman, R. Jackiw, and D. Gross, Princeton University Press, Princeton, N.J. (1972).

[3] K. Johnson, *Phys. Lett.* 5: 253 (1963).

[4] The connection between the axial vector anomaly and mass generation in the Schwinger model was previously established by R. Jackiw in *Laws of Hadronic Matter*, edited by A. Zichichi, Academic Press, New York (1975).

[5] R. Jackiw and S. Templeton, *Phys. Rev. D* 23: 2291 (1981). For further discussion, see W. Siegel, *Nucl. Phys. B* 156: 135 (1979); J. Schonfeld, *Nucl. Phys. B* 185: 157 (1981); S. Deser, R. Jackiw, and S. Templeton, *Phys. Rev. Lett.* 48: 975 (1982) and *Ann. Phys.* (NY) 140: 372 (1982).

[6] The quantization of three-dimensional topologically massive gauge fields has been performed by S. Deser, R. Jackiw, and S. Templeton, ref. [5].

[7] B. Binegar, *J. Math. Phys.* 23: 1151 (1982).

[8] H. Nielsen and P. Olesen, *Nucl. Phys. B* 61: 45 (1973).

[9] R. Jackiw and C. Rebbi, *Phys. Rev. Lett.* 37: 172 (1976). For a review, see R. Jackiw, *Rev. Mod. Phys.* 52: 661 (1980).

[10] For an account, see S. Chern, *Complex Manifolds Without Potential Theory*, 2nd edition, Springer, Berlin (1979). These structures have previously occurred in physical theory in the analysis of four-dimensional Yang-Mills theory; see ref. [9].

[11] R. Jackiw and K. Johnson, *Phys. Rev. D* 8: 2386 (1973); J. Cornwall and R. Norton, *Phys. Rev. D* 8: 3338 (1973).

[12] For a critical review, see E. Farhi and R. Jackiw, *Dynamical Gauge Symmetry Breaking*, World Scientific, Singapore (1982).

Gribov Ambiguities and Quantization in the Axial Gauge

William I. Weisberger

This paper treats some topics in gauge fixing and path-integral quantization of non-Abelian gauge theories. The ambiguities of gauge-fixing procedures noted by Gribov are discussed and resolved by quantization in the axial gauge. Both the Lagrangian formulation using the Faddeev-Popov ansatz and the Hamiltonian formalism with constraints are treated. Attention is paid to the residual gauge invariance which remains after the axial gauge is chosen.

I. Introduction

This paper is intended as a pedagogical treatment of some topics related to gauge fixing and path-integral quantization of non-Abelian gauge theories. It deals mainly with the ambiguities of conventional gauge-fixing procedures first noticed by Gribov [1], and presents a resolution of the problem by quantization in the axial gauge. Gribov was the first to show that conventional gauge fixing via the radiation or Lorentz gauges familiar from the study of electrodynamics does not work in non-Abelian gauge theories. In certain cases there remains a finite multiplicity of gauge-equivalent potentials. Singer [2] generalized Gribov's result for any smooth gauge-fixing function when the vector potential continued to four-dimensional Euclidean space can be compactified on the four-dimensional sphere. This would seem to imply that no global gauge choice is possible. There is a loophole, however. The technical requirement of compactification restricts the choice of gauges. Potentials in the axial gauge cannot be compactified no matter how fast the field strengths vanish at infinity. Therefore, it is natural to adopt the axial gauge as a starting point for quantization. The equivalence of physical results obtained in other gauges can be established for calculations of small quantum fluctuations about classical fields. This includes ordinary perturbation theory, instantons, dilute instanton gases, etc.

After reviewing Gribov's and Singer's results and demonstrating the noncompactifiability of the axial gauge, this paper treats the quantization of non-Abelian gauge theories in the axial gauge by the path-integral formalism. We deal first with the intuitive formulation in the Lagrangian framework using the Faddeev-Popov ansatz [3]. The result obtained is further justified by starting with a Hamiltonian formalism and deriving from it the Lagrangian result. In both cases attention is paid to the residual gauge invariance which remains after the axial gauge is chosen.

II. The Gribov Ambiguity

We start by considering the vacuum-to-vaccuum transition amplitude continued to Euclidean space. For a gauge theory the transcription of the formalism used for a nonsingular Lagrangian would give for such an amplitude

$$Z = \int [\mathcal{D}A] \exp \left\{ -\frac{1}{e^2} \int d^4x \mathcal{L}(A) \right\},$$

$$\mathcal{L} = -\tfrac{1}{8} \mathrm{Tr}\,(F_{\mu\nu} F_{\mu\nu}),$$

(1)

with

$$D_\mu = \partial_\mu + iA_\mu,$$

$$[D_\mu, D_\nu] = iF_{\mu\nu}.$$

The gauge potentials have been scaled by the coupling constant e to simplify notation. Here and henceforth we neglect normalization factors in Z. We are using a matrix notation with

$$A_\mu = I^a A_\mu^a,$$

(2)

where the I^a form a representation of the generators of the Lie algebra associated with the group G of the theory. The generators are normalized to

$$\mathrm{Tr}\, I^a I^b = 2\delta^{ab}.$$

(3)

As is well known, (1) is unsatisfactory because the term in the action quadratic in the vector potential is singular; the corresponding wave operator has no inverse. Therefore, even in the Euclidean formulation there is insufficient Gaussian damping of the integral. This can be ascribed to invariance of the action under the local gauge group $\mathcal{G}$. If $U(g)$ is a

representation of some group element $g \varepsilon \mathcal{G}$, the vector potential transforms under g as

$$A_\mu^g = U(g)A_\mu U^{-1}(g) + i\partial_\mu U(g)U^{-1}(g). \tag{4}$$

As g varies over $\mathcal{G}$, A_μ^g describes an orbit in the space of potentials
$\mathcal{A}$. Since the action is invariant on an orbit, one is led to associate the
singular nature of (1) with the infinite volume $\prod_x dg(x)$ associated with
each orbit. Therefore, one should rewrite (1) as an integral over only the
space of orbits $\mathcal{B} = \mathcal{A}/\mathcal{G}$. The problem then is to introduce appropriate
coordinates and measure for $\mathcal{B}$. Conventional gauge-fixing procedures
start with the implicit assumption that $\mathcal{A}$ is a direct-product space

$$\mathcal{A} = \mathcal{G} \times (\mathcal{A}/\mathcal{G}),$$

so that

$$[\mathcal{D}A_\mu(x)] = [\mathcal{D}B][\prod_x dg(x)].$$

$\mathcal{B}$ is given coordinates by choosing a single representative element on
each orbit by a set of gauge-fixing conditions

$$f^a(A_\mu) = 0. \tag{6}$$

That is, $f^a(A_\mu^g) = 0$ should have a unique solution g for each A_μ. The standard trick to compute the measure is to insert in the measure for (1) the
factor

$$1 = \Delta_f(A_\mu) \int \prod_x dg(x) \prod_{x,a} \delta[f^a(A_\mu^g(x))], \tag{7}$$

which defines the Faddeev-Popov determinant

$$\Delta_f(A_\mu) = \det M_f,$$

with   (8)

$$[M_f(x, y)]^{ab} = \frac{\delta f^a(A(x))}{\delta A_\nu^c(y)} D_\nu^{cb}(A(y)) \bigg|_{f=0}.$$

This leads to the identification

$$[\mathcal{D}\mathcal{B}] = [\mathcal{D}A_\mu]\Delta_f(A_\mu) \prod_{a,x} \delta(f^a(A_\mu(x))). \tag{9}$$

For (9) to be a measure it is necessary that the Fadeev-Popov determinant
be nonnegative.

Gribov was the first to point out that this procedure does not work for

Coulomb and Lorentz gauges in non-Abelian gauge theories. For example, in the Lorentz gauge there are orbits for which

$$\partial_\mu A_\mu^g = 0 \tag{10}$$

has more than one solution for g.

Suppose A_μ and A_μ' are two gauge-equivalent potentials both satisfying the Lorentz gauge condition. There is some local gauge transformation represented by U such that

$$A_\mu' = U A_\mu U^{-1} + i\partial_\mu U U^{-1}. \tag{11}$$

The vanishing of the four-divergence of both A_μ and A_μ' leads to an equation for U, viz.,

$$D_\mu(A_\mu)S_\mu = 0, \quad S_\mu = U^{-1}\partial_\mu U. \tag{12}$$

The development of nontrivial solutions of (12) can be understood by writing a variational principle for (12) in a form attributed to Polyakov,

$$W = \int d^4x \, \mathrm{Tr}\,\{(\partial_\mu U)(\partial_\mu U^\dagger) + 2i(\partial_\mu U)U^\dagger A_\mu\}, \quad U^\dagger U = 1. \tag{13}$$

For $\partial_\mu A_\mu = 0$, W is stationary when U satisfies (12). We can write W as the difference of two positive-definite quadratic forms

$$W = \int d^4x \, \mathrm{Tr}\,\{(D_\mu(A)U)^\dagger(D_\mu(A)U) - (A_\mu U)^\dagger(A_\mu U)\}. \tag{14}$$

$W(U = 1) = 0$, and W is positive definite when $A_\mu = 0$. For the null field the minima is at $U = 1$, and we do not expect to find any copies. From the form of (14), it seems likely that for big enough A_μ we can find trial U's for which W is negative. In that case W must have a minima other than $U = 1$, and a gauge copy exists.

By continuity the gauge copies should emerge smoothly as we vary the potential A_μ. The transformation which takes us to the copy should develop continuously from the identity as we vary A_μ. Therefore we are led to consider infinitesmal transformations of the form

$$U = \exp i\alpha V \approx 1 + i\alpha V, \quad V = V^\dagger, \tag{15}$$

where α is a small real parameter and V is normalized by

$$\int d^4x \, \mathrm{Tr}\,(VV^\dagger) = 1. \tag{16}$$

To quadratic order the variational integral is

$$W \approx \alpha^2 \int d^4x \, \text{Tr} \, \{(\partial_\mu V)(\partial_\mu V) + iV[A_\mu, \partial_\mu V]\}, \tag{17}$$

and the linearized variational equation subject to the constraint (16) is

$$-\partial_\mu D_\mu(A) V = -\partial^2 V + i[A_\mu, \partial_\mu V] = \lambda V. \tag{18}$$

For such an eigenfunction

$$W/\alpha^2 = \lambda \int \text{Tr} \, (V^2) = \lambda. \tag{19}$$

For sufficiently large A_μ we can certainly find a trial V such that the quadratic W is negative. Hence, there is a minimizing V for the linearized problem with $\lambda < 0$. Moreover, if $A_\mu \to 0$ at ∞, then (18) shows that only eigenfunctions with negative eigenvalues λ are normalizable. If the linearized problem has a negative minimum, the unrestricted variation over all U which vanish at infinity must possess a minimum at least as negative as that of the infinitesmal transformations. The minimizing U is the desired transformation.

This leads us to the following picture for the emergence of Gribov copies. Take some form for A_μ in the Lorentz gauge and multiply it by a scale factor η. As η is increased from zero, a second gauge copy arises when $\eta = \eta_0$. The relevant U starts at $U(\eta_0) = 1$ and continuously develops into a finite transformation. As η increases further more gauge copies may be generated.

Furthermore, the differential operator in (18) is just the Faddeev-Popov matrix M. As the first gauge copy emerges, an eigenvalue of the Faddeev-Popov matrix goes through zero and becomes negative. The proposed integration measure of (9) is not acceptable. Two questions naturally arise. How general is the Gribov ambiguity; can it be avoided by choosing another gauge? If not, can we find a gauge copy for which the Faddeev-Popov matrix is positive definite?

The latter question can be answered in the affirmative by considering the positive-definite functional defined over an orbit:[1]

$$I_A(U(g)) = \int \text{Tr} \, (A_\mu^g A_\mu^g) \, d^4x. \tag{20}$$

---

1. I learned the following proof in a seminar by L. Faddeev. He said the result was known to I. M. Singer and others.

Stationary points of $I_A(U)$ give potentials which satisfy the Lorentz gauge. That is,

$$\left.\frac{\delta I}{\delta U}\right|_{U(\bar{g})} = 0 \tag{21}$$

implies

$$\partial_\mu A_\mu^{\bar{g}} = 0 \tag{22}$$

The quadratic fluctuations about such a stationary point obtained by expanding $U = U(\bar{g}) + \phi$ are

$$\Omega(\phi) = \frac{1}{2}\left(\phi, \left.\frac{\delta^2 I}{\delta U \delta U}\phi\right)\right|_{U(\bar{g})} = (\phi', -\partial_\mu D_\mu(A^{\bar{g}})\phi')$$

$$= \int d^4x \, \mathrm{Tr}\,\{\phi'(x)\partial_\mu D_\mu(A^{\bar{g}})\phi'(x)\}, \tag{23}$$

$$\phi' = U(\bar{g})^\dagger \phi.$$

Since I_A is a positive definite operator, it must have at least one minimum as U varies over $\mathcal{G}$. There must be a solution of (21) for which $-\partial_\mu D_\mu(A^{\bar{g}})$ has no negative eigenvalues. If there is a zero eigenvalue with a normalizable eigenfunction ϕ_0, $A_\mu^{\bar{g}}$ defines an orbit on which two Gribov copies are just beginning to split from each other.

The former question mentioned above has been answered in general by Singer [2]. He showed that if the gauge potentials are compactifiable on a four-dimensional sphere S^4, any smooth gauge condition will have multiple solutions on some orbits. If we pick some orbit and a gauge condition which selects a unique representative on that orbit, and then go out along any direction in function space, we will reach an orbit on which the Faddeev-Popov determinant vanishes and a Gribov copy begins to split off. The essence of Singer's proof can be summarized as follows.

Consider the set of all potentials with zero topological charge mapped onto S^4 by stereographic projection.

(1) Suppose it is possible to choose a smooth gauge condition which picks out one and only one representative on each orbit.[2] Then the space of potentials is a direct product space: $\mathcal{A} = \mathcal{G} \times (\mathcal{A}/\mathcal{G})$.

(2) All the homotopy groups of $\mathcal{A}$ are trivial. Any sphere S^n mapped onto $\mathcal{A}$ can be contracted to a point. From (1), $\mathcal{G}$ must be topologically trivial also.

---

2. In fiber bundle language, pick a continuous section of the bundle $\mathcal{A}$ with the fiber $\mathcal{G}$.

(3) However, if $\mathcal{G}$ is non-Abelian, it can be shown to have nontrivial homotopy groups.

(4) Therefore (1) must be false. We cannot fix a global gauge except for $G = U(1)$, i.e., electrodynamics.

This result prevents global gauge fixing. However, we can, in principle, split up the space $\mathcal{A}$ into different regions, each consisting of a set of orbits and fix the gauge locally in each region. Locally, we have the direct product structure for $\mathcal{A}$ and the Jacobian (Faddeev-Popov determinant) is univalent. Therefore the Gribov ambiguity is no impediment to calculating small quantum fluctuations about some classical background field. This includes ordinary perturbation theory and quantum corrections to instanton configurations [4].

However, this is not a completely satisfactory state of affairs if one hopes eventually to understand nonperturbative effects. It does not seem feasible to partition the space of potentials as described above at the start and then carry out the functional integral. The Feynman integral receives its contributions from very jagged discontinuous paths that one can hardly expect to partition in this way.

The gauge-ambiguity problem also raises problems for canonical quantization. If the gauge cannot be fixed ab initio, then one cannot eliminate the constrained variables, define canonical momenta and coordinates, and impose canonical commutation relations [5].

Fortunately, there appears to be a way out of these difficulties. In Singer's general treatment it is essential that the gauge potential be compactifiable on S^4. This requires a rapid fall-off of the potential at ∞ in Euclidean four-space. The rate of decrease depends on the type of gauge chosen, not just on the rate at which the field strengths or Lagrangian density go to zero. Certain classes of gauges, including the axial gauge [6], are intrinsically noncompactifiable and hence may be good gauges [7]. Therefore, we are led to try to define our quantization procedure initially in such a gauge [8].

III. Path-Integral Quantization in the Axial Gauge

A. Noncompactifiability
For the potential to be compactifiable it must be a smooth function when mapped from four-dimensional Euclidean space E^4 onto the four-dimensional sphere S^4 by stereographic projection from the north pole. To study the behavior of the point at ∞, we must remap a neighborhood of the north pole onto a bounded region of E^4. This we can do by inverting the

stereographic projection from the south pole of the sphere. This amounts to a mapping of E^4 onto E^4 by

$$\bar{x}^\mu = x^\mu / x^2, \tag{24}$$

which sends the point at ∞ into the origin. A covariant vector transforms as

$$\bar{A}_\mu(\bar{x}) = \frac{\partial x^\nu}{\partial \bar{x}^\mu} A_\nu(x)$$

$$= \frac{1}{(\bar{x})^2} [\delta^\nu_\mu - 2\hat{\bar{x}}_\mu \hat{\bar{x}}^\nu] A_\nu(x), \tag{25}$$

$$\hat{\bar{x}}^\nu = \bar{x}^\nu / |\bar{x}|.$$

Hence for regularity of $\bar{A}_\mu(\bar{x})$ as $\bar{x} \to 0$, we need

$$\lim_{|x| \to \infty} A_\nu(x) \sim O(1/x^2). \tag{26}$$

Suppose we have a nice smooth potential $A_\mu(x)$ falling rapidly to zero at infinity, so that it is compactifiable. We want to transform to the axial gauge, that is, to find a local gauge function U such that

$$A'_3 = U^\dagger A_3 U + iU^\dagger \partial_3 U = 0 \tag{27}$$

or

$$\partial_3 U = iA_3 U, \tag{28}$$

which has the solution

$$U(x_1, x_2, x_3, x_4) = \left[P_3 \exp i \int_{-\infty}^{x_3} A_3(x_1, x_2, x'_3, x_4) dx'_3 \right] U_-(x_1, x_2, x_4) \tag{29}$$

where P_3 indicates that the integral along the x'_3 axis is path-ordered and $U_-(x_1, x_2, x_4)$ is an arbitrary local group element. U_- manifests the remaining freedom of x_3-independent gauge transformations that exists after fixing $A'_3 = 0$. As $x_3 \to +\infty$, the transformation is determined by

$$U_+(x_1, x_2, x_4) = S(x_1, x_2, x_4) U_-(x_1, x_2, x_4),$$

where

$$S = P_3 \exp i \int_{-\infty}^{\infty} A_3(x) dx_3. \tag{30}$$

We can fix the remaining gauge freedom, for example, by choosing $U_- = 1$. This guarantees that A'_μ vanishes as rapidly as A_μ when $x_3 \to -\infty$. Also, from (29) $U \to 1$ when $|x_1|$, $|x_2|$ or $|x_4| \to \infty$. But when $x_3 \to +\infty$ we have

$$U_+ = S \neq 1.$$

Thus

$$\lim_{x_3 \to +\infty} A'_\mu = iU_+^\dagger \partial_\mu U_+ , \tag{31}$$

which is a pure gauge but does not vanish.

U_+, U_-, and S are all mappings of S^3 onto the group space of G. For a topologically nontrivial potential, the topological charge is given in the axial gauge by the difference between the winding numbers associated with U_+ and U_-, or equivalently, by the winding number of S.

B. Faddeev-Popov Ansatz

From the foregoing one is tempted interpret the Faddeev-Popov ansatz for the vacuum-vacuum functional in the axial gauge as instructing us to write

$$Z = \int [dA_\mu] \prod_{x,a} \delta(A_3^a(x)) \exp -\frac{1}{e^2} \int d^4x \mathscr{L}(A_\mu) \Big|_{U_-} \tag{32}$$

The Faddeev-Popov determinant is field-independent in the axial gauge and may be dropped. The restriction on $U_-(x_1, x_2, x_4)$ that fixes the remaining gauge freedom would be very awkward to implement in practice. One would have to introduce another set of delta functions fixing

$$\lim_{x_3 \to -\infty} A_j = iU_-^\dagger \partial_j U_- , \quad j = 1, 2, 4. \tag{33}$$

However, it seems that we can relax this additional restriction. Since gauge-invariant amplitudes are by definition independent of how we pick a representative on each orbit, we can also define them by averaging over each orbit with some measure which has the same volume on each orbit. A well-known example is the general covariant gauge defined by first fixing

$$\partial_\mu A_\mu^a = f^a(x),$$

and then integrating over all possible gauge-fixing functions f^a with a Gaussian weight

$$\exp\left\{-\frac{1}{2\alpha}\int \sum_a (f^a(x))^2 d^4x\right\}.$$

Since the Faddeev-Popov determinant is independent of $f^a(x)$, the result is to eliminate the gauge-fixing delta functions in favor of adding a gauge-fixing term

$$\frac{1}{2\alpha}\sum_a (\partial_\mu A_\mu^a)^2$$

to the Lagrangian. In the case of the axial gauge it is convenient to average over all choices of U_- with unit weight. The effect is just to drop the restriction on U_- in (32). Thus one is integrating with equal weight over all axial gauge representatives on each orbit.[3] This is a sufficient restriction to render the path integral nonsingular, since the quadratic term in the action can now be inverted to find a propagator.

The prescription of averaging over all axial gauges, which seems a matter of convenience in the preceding discussion, receives support as the required procedure from additional arguments. In the Lagrangian formulation of the path integral, the limiting behavior of A_μ at infinity can be examined by considering the system to be enclosed in a series of large hypercubic boxes of increasing size. Coleman [9] has shown that it is possible to change the boundary conditions on a three-dimensional side as one increases the size of the box without appreciably changing the action integral, as long as the winding number of the potential on that side is not varied. With the limiting behavior defined in this sequential manner, only the winding number survives as relevant physical data. This suggests again that we should relax the restriction on U_- and integrate over all axial gauge potentials.

C. Hamiltonian Formulation

For a dynamical system whose kinetic energy term is coordinate dependent, the path integral should be defined in the Hamiltonian formalism. If the system is characterized by canonical variables p_j and q_j, the vacuum persistence amplitude is written as [10, 11]

$$Z = \int \prod_j [dp_j][dq_j] \exp\left\{\frac{i}{e^2}\int \left[\sum_j p_j \dot{q}_j - H(p,q)\right] dt\right\}. \tag{34}$$

3. To maintain cluster decomposition one should use a θ vacuum. In the axial gauge this can be implemented by adding to the Lagrangian a term $(\theta/2)\varepsilon_{\mu\nu\alpha\beta}\mathrm{Tr}\,(F_{\mu\nu}F_{\alpha\beta})$ (see [12]).

(We have returned to the ordinary spacetime metric for the Hamiltonian treatment.)

For gauge theories, the Lagrangian is singular, and the gauge must be fixed and constraint equations solved before canonical momenta can be introduced and a Hamiltonian derived. Using a first-order formalism

$$\mathscr{L} = \tfrac{1}{8}\operatorname{Tr}(F^{\mu\nu}F_{\mu\nu}) - \tfrac{1}{4}\operatorname{Tr}(F^{\mu\nu}(\partial_\nu A_\mu - \partial_\nu A_\mu + i[A_\mu, A_\nu])), \tag{35}$$

the classical Euler-Lagrange equation are

$$F_{\mu\nu} = \partial_\mu A_\nu - \partial_\nu A_\mu + i[A_\mu, A_\nu],$$
$$D^\mu(A)F_{\mu\nu} = 0. \tag{36}$$

This system is singular because

$$\pi_0^a = \frac{\delta\mathscr{L}}{\delta(\partial_0 A_0^a)} = 0, \tag{37}$$

and the other canonical momenta

$$\pi_j^a = \frac{\delta\mathscr{L}}{\delta(\partial_0 A_j^a)} = F_{0j}^a \equiv E_j^a$$

are not independent, but satisfy the constraint equations

$$D_j E_j = 0 \quad \text{(Gauss's law)}. \tag{38}$$

If we pick the axial gauge $A_3 = 0$, then E_3 and A_0 are given in terms of the independent canonical variables π_i, A_i, $i = x, y$, by solution of the constraint equations

$$E_3 = \nabla_3 A_0,$$
$$\nabla_3 E_3 = -\sum_{j=x,y} D_j \pi_j = -\mathbf{D} \cdot \boldsymbol{\pi} \equiv -\sigma. \tag{39}$$

In solving these equations care must be taken with behavior as $|z| \to \infty$. Therefore we first restrict z to the range $(-L, L)$ and let $L \to \infty$ later. Defining magnetic fields by

$$B_i = -\tfrac{1}{2}\varepsilon_{ijk}F_{jk}, \tag{40}$$

we can write the Hamiltonian density as

$$\mathscr{H}(\pi, A) = \tfrac{1}{2}\operatorname{Tr}(\boldsymbol{\pi} \cdot \dot{\mathbf{A}}) - \mathscr{L}$$
$$= \tfrac{1}{4}\operatorname{Tr}[\boldsymbol{\pi} \cdot \boldsymbol{\pi} + \mathbf{B} \cdot \mathbf{B} - (E_3)^2 - 2\boldsymbol{\pi} \cdot \mathbf{D}A_0]. \tag{41}$$

Integrating by parts with attention to the surface terms gives the Hamiltonian

$$H = \frac{1}{4} \int d^3x \, \mathrm{Tr}\left[\boldsymbol{\pi} \cdot \boldsymbol{\pi} + (E_3)^2 + \mathbf{B} \cdot \mathbf{B}\right]$$

$$- \int d^2x_\perp \, \mathrm{Tr}\left[E_3(x_\perp, z) A_0(x_\perp, z)\right]\Big|_{z=-L}^{z=L}. \tag{42}$$

In order to obtain the familiar positive-definite form for the Hamiltonian, we should choose the surface terms at $z = \pm L$ to cancel each other. This is achieved by the symmetric solutions

$$E_3(x_\perp, z) = -\frac{1}{2} \int \varepsilon(z - z') \sigma(x_\perp, z') dz',$$

$$A_0(x_\perp, z) = \frac{1}{2} \int \varepsilon(z - z') E_3(x_\perp, z') dz' \tag{43}$$

$$= -\frac{1}{2} \int |z - z'| \sigma(x_\perp, z') dz' + \frac{L}{2} \int \sigma(x_\perp, z') dz'.$$

This choice of the Green's function

$$\langle z | \nabla_3^{-1} | z' \rangle = \tfrac{1}{2}\varepsilon(z - z') \tag{44}$$

corresponds to the momentum representation

$$\langle k_3 | \nabla_3^{-1} | k_3' \rangle = -i\delta(k_3 - k_3') \mathrm{P}\left(\frac{1}{k_3}\right) \tag{45}$$

where P denotes principal value—the desired form of the $1/k_3$ singularity in the axial gauge.

When the Hamiltonian is expressed in terms of the independent canonical variables π_i, A_i, $i = x, y$, the Hamiltonian equations of motion

$$\dot{A}_i(x) = \frac{\delta H}{\delta \pi_i(x)}, \quad \dot{\pi}_i(x) = -\frac{\delta H}{\delta A_i(x)} \tag{46}$$

reproduce the Euler-Lagrange equations of motion. Their solution is specified if initial-value data for π_i and A_i are given at some time t_0.

The Hamiltonian retains a residual local gauge invariance under (x, y)-dependent gauge transformations. These remaining gauge degrees of freedom are canonical transformations whose generators are

$$Q^a(x_\perp) = \int \sigma^a(x_\perp, z) dz. \tag{47}$$

The Poisson brackets of H with Q^a vanish. However, the initial-value data is not invariant under these transformations, and fixes the remaining gauge freedom. Sets of initial value data which are related by (x, y)-dependent gauge transformations lead to equivalent physical results.

There is however, a problem when we take the limit $L \to \infty$. The Hamiltonian becomes infinite for [12]

$$\mathrm{Tr} \int d^3x (E_3)^2 = \frac{1}{4} \mathrm{Tr} \int d^3x dz' dz'' \varepsilon(z - z')\sigma(x_\perp, z')\varepsilon(z - z'')\sigma(x_\perp, z'')$$

$$= -\frac{1}{2} \mathrm{Tr} \int d^2x_\perp dz' dz'' \sigma(x_\perp, z')|z' - z''|\sigma(z'') \qquad (48)$$

$$+ \frac{L}{2} \mathrm{Tr} \int d^2x_\perp (Q(x_\perp))^2.$$

Also, the solution for A_0, eq. (43), is singular at $L = \infty$. Therefore, we are led to introduce the additional constraints [13]

$$Q^a(x_\perp) = 0. \qquad (49)$$

These constraints are closed under the Poisson-bracket operation.[4]

$$\{Q^a(x_\perp), Q^b(y_\perp)\} = f^{abc} Q^c(x_\perp)\delta^{(2)}(x_\perp - y_\perp). \qquad (50)$$

Therefore, we can use the procedure of Faddeev [10, 14] to implement them in the path integral. In the subspace of phase space where the constraint (49) is fulfilled, (x, y)–gauge transformations generate curves along which physical observables are invariant. Additional gauge-fixing constraints are required to fix a point along each of these curves. Following Faddeev we want a set of conditions

$$\chi^a(x_\perp) = 0 \qquad (51)$$

that are nonsingular:

$$\det \|\{\chi^a(x_\perp), Q^b(y_\perp)\}\| \neq 0, \qquad (52)$$

and mutually consistent:

$$\{\chi^a(x_\perp), \chi^b(y_\perp)\} = 0. \qquad (53)$$

A convenient choice is

$$\chi^a = A_x^a(x_\perp, z_0) - f^a(x_\perp) = 0 \qquad (54)$$

for some fixed z_0, where the f^a are arbitrary functions.[5] In this case

---

4. The fundamental Poisson bracket is $\{\pi_i^a(\mathbf{x}), A_j^b(\mathbf{y})\} = \delta^{ab}\delta_{ij}\delta^{(3)}(\mathbf{x} - \mathbf{y})$.
5. An apparent remaining freedom of y-dependent gauge transformations is eliminated by the boundary condition $\lim_{|x_\perp| \to \infty} A_i = 0$.

$$\{\chi^a(x_\perp), Q^b(y_\perp)\} = -[D_x(A(x_\perp, z_0))]^{ab}\delta^{(2)}(x_\perp - y_\perp). \tag{55}$$

Then the vacuum persistence functional can be written as

$$Z = \int \prod_{i=x,y} [\mathcal{D}\pi_i][\mathcal{D}A_i] \prod_{a,x_\perp} \delta(Q^a(x_\perp))\delta(A_x^a(x_\perp, z_0) - f^a(x_\perp))$$
$$\times \det \|\{\chi^a, Q^b\}\| \exp\left\{\frac{i}{2e^2} \int d^4x \, \mathrm{Tr}\,(\pi \cdot \dot{A} - \mathscr{H}(\pi, A)))\right\}. \tag{56}$$

Because of the constraint (54), the Faddeev-Popov determinant is field independent. Denote it by $M(f)$.

To return to the Lagrangian formulation from which covariant perturbation rules can be derived easily, we introduce an independent variable E_3 into H by inserting in the integral the factor

$$1 = \int [\mathcal{D}E_3] \prod_{a,x} \delta(E_3^a + \nabla_3^{-1}(\mathbf{D} \cdot \pi)^a)$$
$$\propto \int [\mathcal{D}E_3] \prod_{a,x} \delta(\nabla_3 E_3^a + (\mathbf{D} \cdot \pi)^a). \tag{57}$$

The product of delta functions can be represented as a functional integral by introducing an integration variable we call A_0:

$$\prod_{a,x} \delta(\nabla_3 E_3^a + (\mathbf{D} \cdot \pi)^a) \propto \int [\mathcal{D}A_0] \exp\left\{\frac{-i}{2e^2} \mathrm{Tr} \int d^4x A_0(\nabla_3 E_3 + \mathbf{D} \cdot \pi)\right\}. \tag{58}$$

The proportionality constants are field-independent and can be dropped. Similarly, we exponentiate the Q^a constraint using an integration variable $\lambda^a(x_\perp, t)$. With these substitutions the functional integral becomes

$$Z = \int \prod_i [\mathcal{D}\pi_i][\mathcal{D}A_i][\mathcal{D}E_3][\mathcal{D}A_0][\mathcal{D}\lambda]$$
$$\times \prod_{a,x_\perp,t} \delta(A_x^a(x_\perp, z_0, t) - f^a(x_\perp, t))M(f) \tag{59}$$
$$\times \exp\left\{\frac{i}{2e^2} \mathrm{Tr} \int d^4x [\pi \cdot \dot{A} - \pi \cdot \pi - \mathbf{B} \cdot \mathbf{B} - \nabla_3 E_3 - \right.$$
$$\left. A_0(\nabla_3 E_3 + \mathbf{D} \cdot \pi) - \lambda(x_\perp, t)\mathbf{D} \cdot \pi]\right\}.$$

It is now easy to perform the Gaussian integrals over π and E_3. The argument of the exponential becomes

$$\frac{i}{e^2} \int d^4x \mathscr{L}(A_x, A_y, 0, A_0 + \lambda). \tag{60}$$

Therefore, after a translation,

$$A_0 \to A_0 - \lambda, \tag{61}$$

the λ integral can be absorbed into the normalization factor. This gives

$$Z = \int [\mathscr{D}A_\mu] \prod_{a,x} \delta(A_3^a(x)) \prod_{b,x_\perp,t} \delta(A_x^b(x_\perp, z_0, t) - f^b) M(f)$$
$$\exp\left\{\frac{i}{e^2} \int d^4x \mathscr{L}(A)\right\}. \tag{62}$$

Finally, since gauge-invariant quantities are independent of the choice of f, we can integrate over all functions f with weight $1/M(f)$ to reobtain

$$Z = \int [\mathscr{D}A] \prod_{a,x} \delta(A_3^a(x)) \exp\left\{\frac{i}{e^2} \int d^4x \mathscr{L}(A)\right\}. \tag{63}$$

Green's functions can be generated in the usual way by adding a source term $\int J_\mu A^\mu d^4x$ to the Lagrangian. The equivalence of perturbation results calculated in other gauges can be proved by standard methods. Though the axial gauge is far from the most convenient for such calculations, it does provide a starting point for quantization that is free from Gribov ambiguities and based on a canonical formalism. Though Gribov's gauge-fixing ambiguity is an interesting manifestation of the complicated mathematical structure of non-Abelian gauge theories, it does not appear that it is related in a direct way to confinement or other physical nonperturbative phenomena.

Acknowledgment

This work was supported in part by National Science Foundation grant no. PHY 81-09110.

References

[1] V. N. Gribov, Lecture at 12th Winter School of the Leningrad Nuclear Physics Institute, 1977 (unpublished). (English translation: SLAC-TRANS-176); *Nucl. Phys. B* 139: (1978). See also R. Jackiw, I. Muzinich and C. Rebbi, *Phys. Rev. D* 17: 1576 (1978).

[2] I. M. Singer, *Comm. Math. Phys.* 60: 7 (1978).

[3] L. Faddeev and V. N. Popov, "Perturbation Theory for Gauge Invariant Fields," Kiev ITP Report (unpublished); *Phys. Lett. B* 25: 29 (1967).

[4] D. Amati and A. Rouet, *Phys. Lett. B* 73: 39 (1978).

[5] In the lattice gauge formulation of K. Wilson, *Phys. Rev. D* 10: 2445 (1974), it is not necessary to fix the gauge.

[6] R. H. Arnowitt and S. I. Fickler, *Phys. Rev.* 127: 1821 (1962).

[7] This point is made by Singer in ref. [2].

[8] The derivation of Feynman rules, Ward identities, etc., in the axial gauge has been discussed in many articles. An incomplete list is: E. S. Fradkin and I. Tyutin, *Phys. Rev. D* 2: 2841 (1970); R. N. Mohapatra, *Phys. Rev. D* 5: 2215 (1972); W. Kummer, *Acta Phys. Austriaca* 41: 315 (1975); W. Konetschny and W. Kummer, *Nucl. Phys. B* 100: 106 (1975); *B* 108: 397 (1976); J. Bernstein, *Phys. Rev. D* 15: 2273 (1977).

[9] S. Coleman, "The Uses of Instantons" in *The Whys of Subnuclear Physics*, edited by A. Zichichi, Plenum, New York (1979).

[10] L. Faddeev, *Teor. Mat. Fiz.* 1: 3 (1969) [*Th. Math. Phys.* 1: 1 (1969)].

[11] When the classical Hamiltonian contains terms such as $f(q)p^2$ or $q(q)p$, this prescription corresponds to Weyl ordering of the quantum-operator Hamiltonian. See N. Christ and T. D. Lee, *Phys. Rev. D* 22: 939 (1980), or ch. 19 of T. D. Lee, *Particle Physics and Introduction to Field Theory*, Scientific Press, Beijing (1981).

[12] J. Schwinger, *Phys. Rev.* 130: 402 (1963).

[13] Implementation of the constraint in canonical operator quantization has been discussed by A. Chodos, *Phys. Rev. D* 17: 2634 (1978); I. Bars and F. Green. *Nucl. Phys. B* 142: 157 (1978). Recently, it has been suggested that the residual gauge invariance in the axial gauge results automatically in the vanishing of matrix elements of Q^a between eigenstates of the Hamiltonian with physical polarizations. See A. Hosoya, Y. Kakudo, Y. Taguchi, A. Tanaka and K. Yamamoto, Osaka Univ. preprint OS-GE 81 35 (unpublished).

[14] Modifications of Faddeev's method necessary to apply it to the $A_3 = 0$ constraint have been discussed by S. Kaptanoglu, *Phys. Lett. B* 98: 77 (1981).

Beyond the Mystery of Quantum Mechanics

Adrian Patrascioiu

It is conjectured that in a closed universe charged particles and electromagnetic fields obeying classical dynamics create the stochastic process responsible for the experimentally observed quantum behavior of particles and fields. The major physical consequences of this hypothesis are analyzed. Amongst them are certain (detectable) variations of the atomic constants with the temperature.

As a former student of Francis's I am not very fond of papers without equations. I still remember that the ratio of equations to words is that of signal to noise. As it will be seen, this paper will have hardly any equations. My decision to write it is due to the fact that it addresses an issue which I believe could have important consequences for our understanding of nature. My hypothesis is that quantum behavior does not follow from any new principles, but may arise naturally in a closed universe filled with classical particles and electromagnetic fields, and that this possibility has unmistakable and experimentally verifiable physical consequences [1].

To put the discussion in perspective, I would like to recall what experiments really teach us about quantum behavior. Consider the famous two-slit experiment, and let it be performed with very dim light, such that only one particle goes through the apparatus at a time. Any given photon can touch the screen at any point and there is nothing in the quantum theory to predict its place of arrival. What can be predicted is the histogram which would arise were the experiment performed repeatedly under identical conditions. In my opinion this experiment reveals unmistakably that quantum mechanics is only a statistical theory.

There is nothing unsatisfactory about a statistical description if one understands its physical origin. The so-called theories with hidden variables attempt to produce precisely such an explanation. One reason they have not become very popular is that in the end, at best, they only reproduce the predictions of quantum mechanics without making any new predictions. Moreover, their mathematical formulation does not introduce any computational advantages.

In some physicists' opinion the situation for hidden-variable theories is worse. That is so because of the experimentally observed violation of Bell's inequality [2]. These same experiments have led to speculations that quantum mechanics implies the existence of superluminal connections. Since the hypothesis I would like to advance may be regarded as a theory of "hidden variables," it may be useful to say a few words on this subject.

The prototype experiment proposed by Bell [3] (and performed since) consist of an $s = 0$ state decaying into two spin-$\frac{1}{2}$ particles. The expectation value of the product $(\mathbf{s}_1 \cdot \mathbf{a})(\mathbf{s}_2 \cdot \mathbf{b})$ for arbitrary axes $\mathbf{a}$ and $\mathbf{b}$ is measured. Each individual measurement is performed at two spacelike points. The experimentally detected value is $-\mathbf{a} \cdot \mathbf{b}$, in agreement with quantum mechanics. What does this experiment indicate about the nature of quantum mechanics and the possibility of hidden variables?

First, the experiment tests only statistical predictions. It does so for the special case in which a certain quantity, the total spin s, is known and conserved. Were it not for that it would be impossible for one to conclude from the fact that if s_{1z}, for instance, were found to be $+1$ in this particular measurement, s_{2z} (if measured) would have to be found to be -1. There clearly is nothing superluminal about this situation and one could easily construct similar examples using dice casts and recording only those throws in which, for instance, the sum of the two dice takes a certain value; by watching one die one would know what the other die shows. Moreover, the relative probabilities would change; for instance, if the sum is restricted to be 2, it is a certainty that the throw must be (1, 1). This brings me to the second question: What have we learnt from these experiments about hidden variables? In my opinion this is difficult to answer because it is not clear how to translate the fact that the total spin s is zero into statistical information regarding the spins of the two electrons and the relevant hidden variables in a model-independent manner. Of course any given theory of hidden variables, such as the one I will propose, will have to make a concrete prediction for the outcome of any experiment, which will prove it to be either right or wrong. But I do not feel that the experimental violation of Bell's inequality can be used to rule out from the outset the possibility of quantum behavior arising naturally as a self-sustaining classical stochastic process.

The point of this rather lengthy discussion is to emphasize that quantum mechanics provides only a statistical description of physical processes. Now I would like to indicate a very simple and perhaps inevitable scheme by which such a description may be forced upon us. I will begin with an analogy. Imagine being in a room in which there is a gas at non-zero temperature. Under these conditions, the behavior of macroscopic objects would be accurately described by Newton's laws. Not so for tiny colloidal particles. For sufficiently long times of observation, the only practical way of describing a colloidal particle is statistical. Thus the description by Newton's equations would be replaced by the Einstein-Smoluckowski equation or the Uhlenbeck-Ornstein equation [4]. Incidentally the latter (statistical) description involves not only the properties of the particle (position, velocity, mass, etc.), but also the properties of the medium (temperature, density, etc.).

I would like to suggest that in studying the behavior of microscopic objects such as electrons, protons, etc., we are in a situation similar to the one described above. Indeed, even though we can study our particle in a cavity where there is a very good vacuum, we must remember that the walls of the cavity are not ideal walls, but are themselves made out of atoms. Thus no matter how thick the wall, there will always be some electromagnetic radiation coming from it due to its electrons and protons gravitating around each other. Therefore it seems to me a priori unjustified and very likely wrong to attempt to describe particles as if they were completely isolated and acted upon only by controllable forces.

At this point the reader, unless too outraged to continue, may wonder about the two most famous reasons for introducing the quantum postulates:

(a) the stability of the atom, in the present case the walls of the cavity,

(b) the ultraviolet catastrophy plaguing the black-body radiation.

Beginning with the first question, of course the natural tendency of a (classical) atom would be to collapse. If the universe is closed, the radiation given off cannot escape at infinity, but it could come back and randomly hit other charged particles, making them radiate some more. In this manner a self-sustaining stochastic process may ensue. Its details would clearly depend upon the average energy density in the system.

Is it reasonable to expect that under ordinary circumstances atoms would form and quantum behavior would be observed? Not if the ultraviolet catastrophy mentioned above inevitably occurred, assuming classical dynamics. On the other hand, more than a hundred years after Boltzmann's celebrated hypothesis regarding the equipartition of energy, no rigorous proof exists that the Rayleigh-Jeans distribution follows from the Maxwell-Lorentz equations of motion for the charged particles and the electromagnetic fields. Should such a proof be given, clearly the process I am proposing would fail to stabilize the atom. In the absence of mathematical rigor, the usual arguments for the Rayleigh-Jeans distribution in classical physics have been analyzed and found wanting [5]. It remains for the future, and I am presently studying the problem, to reveal what kind of stable invariant distributions arise from classical dynamics.

In the remaining part of this article I would like to investigate some of the physical consequences of the stochastic process I am proposing. First, it should be clearly energy-dependent and therefore also temperature-dependent. At very high energy densities, the system would behave as a hot, fully ionized plasma. The quantum behavior observed in nature should arise only for sufficiently low energy density. It should be temperature-dependent and in fact if one cooled off the system sufficiently, the atom should start shrinking. Similarly the $g - 2$ of the electron for instance, should change (increase) as the temperature is lowered. Could such a change have escaped detection? A quantitative answer at this time is impossible simply because I cannot solve the Liouville equation and derive the invariant distribution for the particles and the fields. My guess is that the variation of the $g - 2$ of the electron should be about $10^{-10}/^\circ K$ (the inverse of the mass of the electron). Present accuracy measures $g - 2$ to about 10^{-8}, but it is hoped that future experiments will achieve 10^{-10} [6]. Thus whereas a variation of the atomic constants with the temperature may have been undetectable until now, it may be seen in the near future.

Another important consequence is that it would render Planck's constant a phenomenological constant, computable in terms of the average energy density and the mass and charge of the electron and proton. For the universe as a whole, the former can be calculated in terms of Newton's constant of gravitation, Hubble's constant, and the deacceleration parameter to be about 10^{-29} gm/cm^3. Of this amount cosmologists can account for only about 3%, the remainder being the so-called missing mass of the universe [7]. Perhaps the missing mass is really stored in the random electromagnetic fields produced in the universe by classical particles, constantly falling toward each other. Assuming the distribution of the radiation to be

$$u(v) \approx hv^3 e - v/v_0$$

(in order to guarantee Lorentz invariance at sufficiently large distances), $v_0 \approx 10^{13}$ Hz for the universe as a whole would result. Locally both h and

v_0 would change with the temperature, and such a change would be relevant in understanding the early universe or the peculiar behavior of quasars.

Before abandoning this topic it may be useful to notice that the radiation described so far is due mostly to the electron falling towards the proton and depends upon the temperature only weakly, through changes in h and v_0. However, the black-body radiation obeying Planck's law, strongly temperature-dependent, is presumably due to the acceleration of the atom as a whole during collisions (and should clearly depend upon the c. m. velocity, which measures the temperature).

Finally if quantum behavior arises only because of the interaction of the observed particle with the random electromagnetic fields produced by the other particles in the universe, then it pertains only to sufficiently long times of interaction. Just as in the case of Browian motion the description of the process depends upon the length of time of the observation, becoming increasingly classical for shorter and shorter times. Therefore there would be no need to use field theory at very high energies, as it is done in grand unified theories. In this respect, it is to be noticed that all high-precision tests of field theory involve long-time phenomena [8]. On the contrary, at high energies the data are fitted rather well by the tree diagrams of classical field theory.

Some concluding remarks. I have proposed that quantum behavior may not follow from some new principles, but rather be a natural consequence of our inability to observe truly isolated systems; the basic dynamics would be classical, yet the disturbances due to the presence of charged particles in the universe uncontrollable. (I do not think that classical dynamics means Maxwell-Lorentz equations. After all, they are slightly acausal precisely at distances at which pair creation can occur. There is, however, nothing to prevent pair creation in classical physics; presumably then, some suitable modification of classical electrodynamics can be formulated.) The picture I am advancing has two very attractive features for any physical theory: it is intuitive (no ad-hoc hypotheses) and it is experimentally verifiable in the near future. Mathematically it may turn out to have certain similarities with some very interesting approaches to quantum mechanics, such as Nelson's Brownian motion [9] or the random electrodynamics of Marshall, Boyer, Claverie, et al. [10]. The physical picture is quite different though; indeed these theories postulate the existence of certain fields which permeate the universe, are temperature-independent and carry an infinite energy density. In my opinion such assumptions are as hard to comprehend as the existence of

a statistical description of nature without an explanation of what is caus-
ing the fluctuations, which is what quantum theory provides. Moreover,
in the absense of any new physical predictions, such theories, if successful,
could only be regarded as mathematical reformulations of quantum
mechanics.

Acknowledgments

I am grateful to The Institute for Advanced Study for its hospitality, to
the Alfred P. Sloan Foundation for its support, and to B. N. Taylor for
information regarding the $g - 2$ measurements.

References

[1] A more detailed version of this paper appears in A. Patrascioiu, "On the nature of
quantum behavior," I. A. S. preprint, September 1981.

[2] For a discussion of Bell's inequality and its experimental verification see B. d'Espagnat,
Scientific American, November 1979, p. 128.

[3] J. S. Bell, *Physics* 1: 195 (1964).

[4] See N. Wax, "*Selected Papers on Noise and Stochastic Processes*," Dover, New York
(1954).

[5] A. Patrascioiu, "On the equipartition of energy," I. A. S. preprint, September 1981.

[6] H. Dehmelt in *Atomic Physics*, vol. 7, edited by D. Kleppner and F. M. Pipkin, Plenum,
New York and London (1981).

[7] S. Weinberg, *Gravitation and Cosmology*, Wiley, New York (1972).

[8] T. Kinoshita, Cornell Univ. preprint, 1981.

[9] E. Nelson, *Phys. Rev.* 150: 1079 (1966).

[10] See, e.g., T. H. Boyer, *Phys. Rev. D* 11: 790 (1975).

The Simple Facts about the Baryon Asymmetry of the Universe

The fact that the universe exists and contains matter, such as ourselves, is something most of us are happy to take for granted without questioning how this matter came into existence.

Recently, however, advances in theoretical physics have brought us to a point where it is possible to estimate the relative amounts of energy in the universe present in the form of matter (baryons) to the energy present in the form of radiation (black-body photons). The ratio of baryon number to photon number is presently empirically determined [1] to be of order 10^{-10}, and is crying for an explanation. Let's try to see how this could come about. First we need to know something about the universe. If N is the number of helicity states of all particle species (typically $N \approx 10^2$), the expansion rate of the universe is given by [2]

$$\frac{\dot{R}}{R} \equiv H = 1.66(T)^2 \frac{N^{1/2}}{M_p}, \tag{1}$$

where T is the temperature of the universe in units where $k = 1$ and M_p is the Planck mass $M_p \approx 10^{19}$ Gev. We see therefore that baryons, with a characteristic mass of 1 Gev, will be in thermal equilibrium with photons until the temperature of the universe falls below ≈ 1 Gev, i.e., up until the age of the universe H^{-1} is of order 10^{-7} sec. As the universe subsequently cools, the baryons annihilate with the antibaryons, leaving as residue whatever asymmetry was present between their numbers.

The scenario [3] we will be discussing is one in which we have a grand unified theory (GUT) so that baryon number is not conserved, though the particles, generically labeled X, which mediate baryon-number-violating forces, are extraordinarily heavy (typical mass $\approx 10^{14}$ Gev). We assume then that whatever the initial baryon number of the universe was, it had relaxed to zero through the action of these baryon-number-violating forces by the time $T \approx 10^{14}$ Gev, i.e., $H^{-1} \approx 10^{-34}$ sec. At about this time T fell below the mass of the superheavy particles X, so they began to depart from the thermal equilibrium that we assume they had reached. They simultaneously began to decay into lighter particles such as quarks and leptons because their decay width (the denominator in (2) is due to time dilation) is

$$\Gamma_X \approx \frac{\alpha_X M_X^2 N}{[T^2 + M_X^2]^{1/2}} \approx M_X, \tag{2}$$

for $\alpha_X \approx 10^{-2}$ typical of GUT theories; i.e., the play starts when T, M_X, and Γ_X are all of the same order of magnitude.

Now assume X decays into two channels with baryon numbers B_1 and B_2 at rates r and $1 - r$, respectively. The antiparticle of the X, the $\bar{X}$, decays into two channels with baryon numbers $-B_1$, $-B_2$ at rates $\bar{r}$ and $1 - \bar{r}$. Now the X and $\bar{X}$ are present in equal numbers at thermal equilibrium by TCP (we are assuming it holds), since $M_X = M_{\bar{X}}$ and the distribution $\propto e^{-M/kT}$. Furthermore $\Gamma_X = \Gamma_{\bar{X}}$ by TCP so we have normalized both to one. CP violation, however, allows r to be different from $\bar{r}$. Consequently and X and $\bar{X}$ decay leads to a baryon number

$$\begin{aligned} \Delta B &= \{rB_1 + (1 - r)B_2 - \bar{r}B_1 - (1 - \bar{r})B_2\} \\ &= (r - \bar{r})(B_1 - B_2). \end{aligned} \tag{3}$$

Since typically $B_1 - B_2$ is of order one, we see that all we are left with is the calculation of $\varepsilon = r - \bar{r}$.

Recapitulating, the necessary ingredients for a calculable nonzero ΔB are: (i) baryon-number-violating interactions, (ii) CP and C violation, and (iii) departures from thermal equilibrium for the X particles before decaying. In addition, of course, we assume that the universe reached a state of thermal equilibrium in which whatever baryon number was present at $t = 0$ has been erased.

This essay will continue in the form of a series of questions, some naive and some not so naive, with brief accompanying answers. A few are questions I've asked myself, but most have been posed by people interested in this fascinating problem. Without further ado, let's begin.

Q.1) You are supposedly calculating n_B/n_γ. How do you know what n_B is?

A.1) Well, really I mean $(n_B - n_{\bar{B}})/n_\gamma$; i.e., I'm calculating the asymmetry between baryons and antibaryons. There's an uncertainty of a couple of orders of magnitude in the determination of n_B; bounds on the low side are provided by galactic masses and on the high side by the critical mass necessary to close the universe [4].

Q.2) Well, what about $n_{\bar{B}}$? Couldn't we have $n_{\bar{B}} = n_B$?

A.2) Protons and antiprotons annihilate, of course, to form high-energy particles. Data from gamma rays and from neutrinos show that there is no appreciable mixing of matter and antimatter on as small a scale as galactic clusters [5]. There are some antiprotons present in cosmic rays, but these are expected to be there at a level of about 10^{-4} relative to protons due just to cosmic ray protons colliding with the interstellar medium. There are some anomalies in the recent observations [6], in particular a larger $\bar{p}$ flux than expected from cosmic rays, but it seems unlikely that they originate from antigalaxies. If there were clusters of galaxies and of antigalaxies arranged so that $n_B - n_{\bar{B}} = 0$, one would have to provide an explanation of how they came to be separated. I recommend to you the review article by Steigman on this subject [5].

Q.3) You told me about n_B and $n_{\bar{B}}$, but what is n_γ? How do I know how many photons there are? By the way, I sometimes see the baryon asymmetry of the universe quoted as the ratio of $n_B - n_{\bar{B}}$ to the specific entropy s. Why is that done?

A.3) One way of saying the essential point is that we assume the baryon asymmetry is created very early in the universe and the number of baryons doesn't change after that. The baryon *density*, however, decreases like $1/R^3$ because the universe is expanding; since [2] $R \propto 1/T$, we find $n_B - n_{\bar{B}}$, which we will call n from now on, goes like T^3. For relativistic particles, the energy density ρ goes like T^4, the number density like T^3; in particular at present essentially all the photons in the universe are in the black-body radiation at $T_\gamma = 2.9°$K and

$$n_\gamma = 2 \frac{\zeta(3)}{\pi^2} T_\gamma^3. \tag{4}$$

The entropy density for relativistic particles is $s = 4\rho/3T$, so s also $\propto T^3$. At present s is due to the black-body photons already mentioned and the three (at least) species of neutrinos v_e, v_μ, v_τ. We find for the present universe

$$s = \frac{2\pi^2}{45} \left[2 + 6 \left(\frac{7}{8} \right) \frac{4}{11} \right] T_\gamma^3, \tag{5}$$

the neutrinos being at a lower temperature $T_v = (\frac{4}{11})^{1/3} T_\gamma$ than the photons, since neutrinos decoupled before the period of $e^+ e^-$ annihilation, which heated the photons. Comparing (5) and (4) we obtain the relationship $s \approx 7n_\gamma$.

The key assumption now is that the expansion of the universe has been isentropic, i.e., entropy has been conserved and entropy density of course $\propto T^3$. Therefore n/s has been constant ever since the epoch when the baryon-number-violating interactions ceased to be effective, namely when T fell appreciably below $M_X \approx 10^{14}$ Gev.

Q.4) Couldn't it be that something happened to generate entropy in the universe after the era of baryosynthesis, so that measuring n/s now doesn't tell us what n/s equaled when $T \approx 10^{14}$ Gev; i.e., n/s hasn't been constant?

A.4) Absolutely. As an example, there has recently been a great deal of work done on the problem of phase transitions in the early universe. We believe that at least three phase transitions took place: the first was when SU_5 (or whatever your favorite GUT is) broke to $SU_3 \times SU_2 \times U_1$; the second when $SU_2 \times U_1$ broke to $U_{1\,em}$; and the third when the quark—antiquark condensate formed, leading to the physical proton and neutron masses. The limits on n/s typically say that the entropy of the universe cannot have been increased by more than a factor of $\sim 10^7$ in these phase transitions [7] (more about this later). This spoils a possible solution to the monopole problem since n_{monopole}/s could be made small by increasing s.

Q.5) I'm still worried by your statement that any initial baryon asymmetry in the universe was erased by the time the X bosons decayed. Is that obvious?

A.5.) No, it's not. Another way of phrasing the question is to ask: How soon could the universe have reached thermal equilibrium? At that time, the baryon number would be zero. Ellis and Steigman [8] argue that the horizon of the universe was too small for equilibrium to have been reached when $T \approx M_P \approx 10^{19}$ Gev. At these high temperatures, decay processes are unimportant, and scattering and production processes are estimated. Ellis and Steigman conclude that thermal equilibrium could have been reached by the time $T \approx 10^{16}$ Gev, but probably not much earlier. This adds one more uncertainty to our calculations; but if the X bosons don't begin to effectively decay until $T \approx M_X \approx 10^{14}$ Gev, our picture of baryosynthesis originating in X decays still holds.

Q.5) You listed the three conditions for generating a baryon asymmetry as (i) baryon-number-violating interactions, (ii) CP violation, and (iii) departure from thermal equilibrium. The first is clear, and you did more or less explain (ii) and (iii) after formula (2), but would please spell it out a little better?

A.5) The problem is stated particularly clearly by Kolb and Wolfram [9]. Following their argument, let $S_{if} = (1 + iT)_{if}$ be the amplitude for a transition from state i to state f. The CP conjugate states are $\bar{i}$, $\bar{f}$, and CPT requires $T_{if} = T_{\bar{f}\bar{i}}$ and, of course, that the masses and lifetimes of single-particle CP conjugate states be equal. By unitarity, $\sum_f |S_{if}|^2 = \sum_f |S_{fi}|^2 = 1$; using TCP we then have

$$\sum_f |T_{if}|^2 = \sum_f |T_{\bar{i}f}|^2 = \sum_f |T_{f\bar{i}}|^2 = \sum_f |T_{fi}|^2, \tag{6}$$

where the sum over f includes both states and their antistates. In thermal equilibrium, all states of a system with a given energy are equally populated as long as there is no chemical potential that fixes a quantum number (in this case baryon number); or, alternatively stated, the density of states is only a function of the energy. We then see from the last equality in (6) that i and $\bar{i}$ are produced in equal numbers, so no net baryon number is generated.

Q.7) Are there any other general requirements for generating a baryon asymmetry?

A.7) Yes, there is one important requirement, which I haven't listed, namely that in the $i \rightarrow f$ transition there be an energetically allowed physical intermediate state. In terms of Feynman diagrams this means that T_{if} has a nonvanishing discontinuity. To see why this is necessary, note that according to CPT, $T_{fi} = T_{\bar{i}\bar{f}}$ (sometimes the tilde is put on the $T_{\bar{i}\bar{f}}$ to remind the reader that the spin of particles also change under PT; but we will omit this nicety and assume throughout that we are summing and averaging over spins). Unitarity tells us that

$$i(T_{if} - T_{if}{}^\dagger) = i(T_{if} - T_{fi}{}^*) = \sum_n T_{in} T_{nf}{}^\dagger, \tag{7}$$

but if there are no real intermediate states, the r.h.s. of (7) is zero, so, using TCP,

$$T_{if} = T_{fi}{}^* = T_{\bar{i}\bar{f}}{}^*, \tag{8}$$

which implies

$$\Gamma_{if} = |T_{if}|^2 = |T_{\bar{i}\bar{f}}|^2 = \Gamma_{\bar{i}\bar{f}}. \tag{9}$$

We knew by TCP that the total transition rate for i and $\bar{i}$ were equal (i.e., the lifetime of X equals that of $\bar{X}$), but (9) says the partial rates are equal *in tree approximation* (when the l.h.s. of (7) equals zero), i.e., $\varepsilon = r - \bar{r} = 0$. We thus generate no baryon asymmetry [9, 10, 11].

A second requirement, proved by Nanopoulos and Weinberg [11], is that no baryon asymmetry is generated in X decays to first order in the baryon-number-violating interaction, whose Hamiltonian we call H'. To see this, let's calculate to first order in H', but all orders in other interactions:

$$T_{Xf} = (\psi_f^{\text{out}}, H' \psi_X), \tag{10a}$$

$$T_{\bar{X}\bar{f}} = (\psi_{\bar{f}}^{\text{out}}, H' \psi_{\bar{X}}) = (\psi_X, H' \psi_f^{\text{in}}), \tag{10b}$$

where TCP has been used in (10b). Inserting a complete set of intermediate states n we have

$$T_{\bar{X}\bar{f}} = \sum_n (\psi_X, H' \psi_n^{\text{out}})(\psi_n^{\text{out}}, \psi_f^{\text{in}})$$

$$= \sum_n (\psi_X, H' \psi_n^{\text{out}}) S_{nf}^0, \tag{11}$$

where S_{nf}^0 is the S matrix in the absence of baryon-number-violating interactions. The total decay rate of $\bar{X}$ to all states $\bar{f}$ with a given value $-B$ of the baryon number is then

$$\bar{\Gamma}(-B) = \sum_{f:B(\bar{f})=-B} \rho_{\bar{f}} |T_{\bar{X}\bar{f}}|^2, \tag{12}$$

where $\rho_{\bar{f}}$ is a phase-space factor. By TCP, $\rho_{\bar{f}} = \rho_f$, and by unitarity,

$$\sum_{f:B(f)=B} \rho_f S_{nf}^0 S_{mf}^0{}^* = \begin{cases} \rho_n \delta_{nm} & : B(n) = B \\ 0 & : B(n) \neq B \end{cases}. \tag{13}$$

This immediately allows us to conclude that $\Gamma(B) = \bar{\Gamma}(-B)$, as can be seen by inserting the value of $T_{\bar{X}\bar{f}}$ derived in (11) into (12).

Q.8) Now that the preliminaries are over, how do you calculate n/s in a particular model, say the simplest SU_5?

A.8) I'll assume you know a certain amount about SU_5 [12], in particular that each fermion family (e.g., u, d, e, v) is placed in a left-handed 10-representation (let's just shorten this to rep). and a right-handed 5-rep. We'll call them $\chi_{\alpha,\beta \, \text{L}}$ and $\psi_{\alpha \, \text{R}}$. The gauge bosons $A_{\mu\alpha}{}^\beta$ are in the adjoint 24-rep. and the Higgs bosons are in the adjoint and fundamental reps. We'll call them $\Phi_\alpha{}^\beta$ and ϕ_α: of course only ϕ_α couples to fermions since the fermion bilnears 10×10, 5×5, and $10 \times \bar{5}$ couple to Higgs mesons of the 5 but not the 24. The five ϕ_α fields consist of a superheavy color triplet, flavor singlet and a color singlet, flavor doublet which acts like the ordinary $SU_2 \times U_1$ Higgs doublet. The color-triplet bosons are

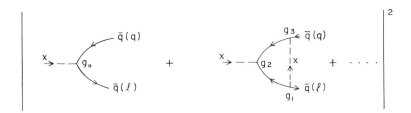

Figure 1
X-boson decay.

candidates for the desired X bosons. They have a mass and decay rate of the scale at which SU_5 breaks to $SU_3 \times SU_2 \times U_1$, namely, $M_X \approx 10^{14}$ Gev and Γ_X given by equation (2). They couple both quarks to quarks and quarks to leptons so the X couplings do not conserve baryon number; their coupling constants are arbitrary complex numbers so they violate CP. Finally, the third of the big three requirements, namely departure from thermal equilibrium, can be satisfied as follows: the X bosons begin to decay when the lifetime of the universe has reached their (decay rate)$^{-1}$ or, using (1) and (2), they decay at a temperature T_{decay} given by

$$\Gamma_X \approx H \Rightarrow T_{\text{decay}} \approx (N^{1/2}\alpha_X M_X M_P)^{1/2}. \tag{14}$$

The X bosons are out of equilibrium when they decay if $M_X \gtrsim T_{\text{decay}}$, i.e., if

$$M_X \gtrsim N^{1/2}\alpha_X M_P, \tag{15}$$

which, with $M_X \approx 10^{14}$ Gev, requires $\alpha_X \lesssim 10^{-6}$, a reasonable value for a Higgs boson-coupling constant. The other conditions, which I discussed in the answer to question seven, are met by carrying the X-decay matrix element to one loop, as shown in figure 1. The one-loop diagram has a real discontinuity since $X \to \bar{q}q$ or ql (q is a quark and l a lepton) and by having an X exchanged, we are going beyond first order in baryon-number-violating couplings.

The dominant contribution to n/s is estimated from the interference of the Born diagram and one-loop diagram as calculated for $X \to \bar{q}\bar{q}$ ($B = -\frac{2}{3}$) and $X \to ql$ ($B = \frac{1}{3}$). The same is done for $\bar{X} \to qq$, $\bar{q}l$ and the difference of the rates multiplied by the relative baryon number, normalized to the overall X rate, gives ΔB of (3). Schematically, if we call g_0 the Born coupling and $g_1 g_2 g_3 I$ the one-loop diagram, we obtain [11, 13] (ω is a phase-space integral)

$$\varepsilon \approx (r - \bar{r}) \approx \frac{1}{|g_0|^2} \left\{ \int d\omega |g_0 + g_1 g_2 g_3 I|^2 \right.$$

$$\left. - \int d\omega |g_0^* + g_1^* g_2^* g_3^* I|^2 \right\} \quad (16)$$

$$= \frac{1}{|g_0|^2} 4 \int d\omega \, \mathrm{Im} \, (g_0^* g_1 g_2 g_3) \, \mathrm{Im} \, I.$$

In evaluating (16) we of course always neglect fermion masses since they are in the Gev range, as compared to $M_X \approx 10^{14}$ Gev (in fact, technically speaking the fermions are all massless at this stage since $SU_2 \times U_1$ is not broken until $T \lesssim 10^2$ Gev, at which point the neutral flavor doublet Higgs boson acquires a vacuum expectation value).

Q.9) Well, does it work for SU_5, i.e., do you get the right n/s?

A.9) The answer isn't simple. To see why, we need to calculate the coupling constants g_i. The Yukawa interaction of fermions and Higgs bosons, for K 5-reps of Higgs bosons, is

$$\mathscr{L}_{\mathrm{Yukawa}} = \sum_{k=1}^{K} \{ \tfrac{1}{2} \bar{\chi}^{\alpha\beta} M_k (\phi_{\alpha'k} \psi_\beta - \phi_{\beta,k} \psi_\alpha)$$

$$- \tfrac{1}{4} \varepsilon^{\alpha\beta\gamma\delta\tau} \chi_{\alpha\beta} C^\dagger M_k' \chi_{\gamma\delta} \phi_\tau \}, \quad (17)$$

where C is the charge-conjugation matrix and where we have omitted the R, L handedness subscripts on the fermion fields as well as the family subscripts. For N families, χ and ψ are each to be thought of as N-component vectors in family space and M and M' are arbitrary $N \times N$ complex matrices. In addition, if we have K 5-reps of Higgs bosons, there are K independent $N \times N$ M and M' matrices, each coupling to a given ϕ_k. Minimal SU_5 says $K = 1$, i.e., we have only one 5-rep of Higgs bosons. In that case, summing over all families in (16) gives

$$\mathrm{Im} \, (g_0^* g_1 g_2 g_3) = \mathrm{Im} \, \mathrm{Tr}(M M^\dagger M'^\dagger M') = 0. \quad (18)$$

since $M M^\dagger$ and $M'^\dagger M'$ are both Hermitian matrices and the imaginary part of the trace of the product of two Hermitian matrices is zero. With a little more work, one can show that the two-loop diagram contribution to ΔB also vanishes, so the first nonzero contribution comes from the interference of the Born term with the three-loop diagram [11, 13]. This means that

$$\varepsilon \lesssim \left(\frac{g_{\mathrm{eff}}^2}{4\pi} \right)^3 \approx \left(\frac{\sqrt{2} G_F m_{\mathrm{eff}}^2}{2\pi} \right)^3 \times \frac{1}{3^6}, \quad (19)$$

where g_{eff} and m_{eff} are some average values of the Higgs couplings and fermion masses. Formula (19) can be somewhat misleading, since at the time of baryosynthesis the fermion masses are all zero because $SU_2 \times U_1$ has not yet been broken. We are of course only using the known low-energy phenomenology to calculate the Higgs coupling constants, i.e., for a fermion f we have $g_f = m_f \sqrt{2}/v$, and $\frac{1}{2}v^2 = G_F/\sqrt{2}$, with v being the vacuum expectation value of the neutral Higgs doublet field. The factor of $(3)^6$ is an approximate correction for the variation of the Higgs couplings from a scale $q^2 \approx (1 \text{ Gev})^2$, where the identification with the fermion mass is made, to $q^2 \approx (10^{14} \text{ Gev})^2$ where baryosynthesis supposedly occurs [14]. Even with $m_{\text{eff}} \approx 10$ Gev, we see $\varepsilon \approx 10^{-12}$, which is clearly too small; we say clearly because there are several factors we have neglected, such as the need for CP violating phases δ_{CP} so $\varepsilon \propto \sin \delta_{CP}$, the need for mixing angles (with only one family $\varepsilon = 0$ as the appropriate traces of M's and M'''s are real), the fact that all particles contribute to the entropy but only X bosons to $n_B - \bar{n}_B$, etc. A more realistic estimate gives

$$\frac{n}{s} \lesssim 10^{-18}. \tag{20}$$

One could try to make g_{eff} bigger by introducing, e.g., a fourth and fifth family of fermions with masses $\gtrsim 10^2$ Gev. This could conceivably work, but it requires a good deal of stretching of limits [15]. It might also cause some trouble with our basic scenario since we now have $\alpha_X \approx 10^{-1}$, whereas in (15) we assumed $\alpha_X \lesssim 10^{-6}$.

A simpler solution is to introduce another Higgs multiplet coupling to fermions, e.g., a second 5-rep, so that instead of (18), we have

$$\varepsilon \propto \text{Im Tr}(M_1 M_2{}^\dagger M_1'{}^\dagger M_2'), \tag{23}$$

where the subscript denotes the Higgs multiplet to which the fermions couple. In this case the one-loop diagram for X_1 decay involves X_2 exchange; the result is that $\varepsilon \lesssim g_{\text{eff}}^2/4\pi$ instead of $\varepsilon \lesssim (g_{\text{eff}}^2/4\pi)^3$ so a reasonable value of n/s is obtainable.

Q.10) If I understood you correctly, you said the prediction for n/s is too small for SU_5 with a single 5-rep, but not too small if a second 5-rep is introduced. That's all right, however, because we probably want a second Higgs doublet in our theory for a variety of reasons, e.g., to solve the strong-CP problem by a Peccei-Quinn U_1 [16], or if we start with a supersymmetric theory. Am I wrong again?

A.10) Unfortunately, you may be. The cases you alluded to above do have two Higgs doublets (or two 5 reps if we extend to SU_5), but their quark couplings are of a specific sort: one doublet couples only to charge-$\frac{2}{3}$ right-handed quarks and the other to charge-$-\frac{1}{3}$ right-handed quarks. This ensures flavor conservation of neutral Higgs boson couplings (i.e., the unitary transformations which diagonalize the mass matrix and hence lead to the physical quark eigenstates also diagonalize the neutral Higgs boson couplings since the two are identical), which is good, but gives a zero n/s in the one-loop approximation. The simple way to see this is to note that we have in this model, from (17)

$$M_1 \neq 0 \quad M_2' \neq 0 \; ; \; M_2 = 0 \quad M_1' = 0$$

or   (24)

$$M_2 \neq 0 \quad M_1' \neq 0 \; ; \; M_1 = 0 \quad M_2' = 0.$$

In either case ε to one loop, as given by (23), is zero. The naive ε estimated in this model is even smaller than the ε calculated using one 5-rep, which couples to all quarks. Of course the real ε may be a lot bigger than the naive ε, as doubtless many clever papers will soon be showing.

Q.11) To tell you the truth, I never liked Higgs bosons. How about using the decay of superheavy gauge bosons? Does that work any better?

A.11) My answers are getting monotonous, in that I keep saying unfortunately no. The main problem is that gauge couplings to fermions are real. Dropping GUT indices, but keeping family indices, they are of the form

$$g(\bar{\psi}_i \gamma_\mu \psi_i A^\mu + \bar{\chi}_i \gamma_\mu \chi_i A^\mu),$$ (25)

so instead of the complex $N \times N$ matrices M, M' that characterized Higgs bosons, we have a real $N \times N$ matrix. Gauge-boson decays therefore automatically have a factor of $g^2/4\pi \approx 10^{-2}$ that Higgs boson decays do not have, and by themselves cannot provide the complex couplings that give CP violation. You might be wondering what I'm saying, because you know that, e.g., the W couplings of quarks to quarks can be complex, via the phases in the Kobayashi-Maskawa model. These arise, however, when we transform from the quark fields p_i, n_i that enter in (25) to the mass eigenstates u_i, d_i by unitary transformations (i is the family index) U', U:

$$\bar{n}_i \gamma_\mu (1 + \gamma_5) p_i W^\mu = \bar{d}_i (U^\dagger U')_{ij} \gamma_\mu (1 + \gamma_5) u_j W^\mu.$$ (26)

$U^\dagger U'$ is the K-M matrix. At the time of baryosynthesis $SU_2 \times U_1$ has not been broken, the quarks are massless, and we can just as well do all our calculations in terms of p_i and n_i fields. These do have complex CP-violating couplings to Higgs bosons, but not to gauge bosons.

In addition, we anticipate that gauge bosons will be heavier and more strongly coupled than Higgs bosons and therefore less able to generate a baryon asymmetry.

Q.12) So SU_5 has problems. Are things better for another GUT? There's a lot of talk about, e.g., SO_{10}.

A.12) In fact it's easier to generate a baryon symmetry in SU_5 than in SO_{10}. Stated more generally, the unitary groups are the most favorable candidates for a nonzero n/s. The reason is connected to the fact that one can generally define a group charge-conjugation operator C [17]; if the symmetry associated with this operator is unbroken, n/s vanishes. We therefore anticipate a suppression of n/s by a factor of Λ_C^2/M_X^2, where Λ_C is the scale at which C is broken and M_X is the scale at which baryo-synthesis occurs. To understand this, consider the decay $X \to ff'$, where the final two fermion state has a certain baryon number B. Now C is an element in the GUT symmetry group G and therefore transforms members of an irreducible representation of G onto themselves; this implies that $C|X\rangle = |X^C\rangle$, $C|f\rangle = |f^C\rangle$. The contribution to baryon asymmetry of $X \to ff'$ is canceled by that of $X^C \to f^C f'^C$ since the $f^C f'^C$ state has baryon number $-B$ and the rates are equal by G invariance.

This problem is avoided in SU_N theories since there is no C operator belonging to G; as an illustration from SU_5, d_L belongs to the 10-rep while $d_L{}^C$ belongs to 5-rep, i.e., d_L and $d_L{}^C$ belong to different irreducible representations. In SO_{10}, on the other hand, the spinorial 16-rep consists of u_L, $u_L{}^C$, d_L, $d_L{}^C$, e_L^+, e_L^{+C}, v_L, $N_L{}^C$. We label the final fermion as $N_L{}^C$ rather than $v_L{}^C$, since we anticipate it being a Majorana spinor. In this case the breaking of C gives a large mass to N and none to v by, e.g., breaking SO_{10} down to $SU_5 \times U_1$. We anticipate the suppression of n/s to be of order M_N^2/M_X^2, which is intolerable unless we have $M_N \approx M_X$ and hence a very large breaking of the left-right symmetry in the decomposition of $SO_{10} \to SU_4 \times SU_{2L} \times SU_{2R}$ [18].

Q.13) This has been a little depressing. I trust it is possible to generate a $n/s \approx 10^{-10}$ in a GUT. You agree, don't you?

A.13) Certainly! I've only been pointing out that it's not straight-forward. If you believe in the general scenario for baryosynthesis, which you must admit is quite attractive and plausible, generation of the correct asymmetry places a very nontrivial constraint on your GUT and also on the evolution of the universe after baryosynthesis. There are a lot of ingenious people interested in this problem, who have found ways to fix all these problems. I'm naturally most familiar with my own work and have cooked up solutions for the SU_5 with two 5-reps with the additional U_1 symmetry, for the minimal model, for theories with a stable proton, etc. Other people have ingenious schemes such as Majorana neutrino decays, symmetries which are broken at high temperatures, and so on. I would say that none of these schemes is compelling, but that's probably just a reflection of the fact that we don't yet have a satisfactory understanding of the mechanisms responsible for symmetry breaking.

Q.14) Is it possible to predict the value of ε and hence n/s from accelerator experiments? I don't mean necessarily present experiments, but conceivable ones.

A.14) Our best bet is the minimal SU_5 model with only one 5-rep of Higgs, since the vacuum expectation value of the neutral component of the doublet is fixed by the W mass and the Higgs couplings appear to be fixed by the fermion masses. We saw in question nine that this model had serious problems, but suppose we could overcome these by introducing a fourth and fifth family of fermions [15] with massive quarks t', b', t'', b''. We could conceivably then measure the K-M matrix of all quarks by studying in detail their $W^{\pm}$ couplings. This determines the unitary transformations between the weak eigenstates p_i, n_i and the mass eigenstates u_i, d_i. It turns out however that the color-triplet Higgs bosons, whose decays are responsible for baryosynthesis, have couplings determined only up to phases; to be precise, for N families there are $N-1$ phases which are unknown and unmeasurable by conceivable experiments. All other models have more free parameters [19]. It is of course conceivable that in some future theory we will be able to predict n/s, but for the moment it is only a constraint on GUT's.

Q.15) I think I've heard enough about GUT's for awhile. Would you please briefly give me a feeling for the dynamics of the situation? I understand X-boson decay generates a baryon asymmetry, but isn't it partially erased by baryon-number-violating forces? How big of an effect is this?

A.15) It's possible to give a feeling for what goes on. As Turner [20] has repeatedly emphasized, it is convenient to define a parameter K (H and Γ_X are defined in (1) and (2)):

$$K = \frac{\Gamma_X}{H}\bigg|_{T=M_x} \approx \frac{3 \times 10^{17}\alpha_X}{M_X}\text{Gev.} \tag{27}$$

Let's separate the discussion into scenarios where $K < 1$ and $K > 1$. Of course we are oversimplifying, but the conclusions are in fact verified by detailed numerical calculations. For $K < 1$ we see that the expansion rate of the universe exceeds the X-boson decay rate for $T = M_X$. As T drops and the X bosons go out of equilibrium, they still remain as abundant as photons even though $T < M_X$ and eventually they decay when $\Gamma_X \approx H$. The resulting $n/s \approx \varepsilon/N \approx 10^{-2}\varepsilon$; subsequent baryon-number-violating exchange forces do not appreciably erase this asymmetry. For instance, the collision rate Γ_c due to X exchange is

$$\Gamma_c \approx (\alpha_X)^2 \frac{T^5 N}{[T^2 + M_X^2]^2}, \tag{28}$$

and since the X-particles decay when $T < M_X$, we see the $(T/M_X)^4$ suppression factor in Γ_c makes it ineffective as an eraser of asymmetry.

For $K > 1$ the results are surprising in that one naively would expect the asymmetry to be erased, since the decay rate is faster than the expansion rate. Detailed calculations show this to not be the case; they conclude that n/s only falls with K like $\sim (1/K)^{1.3}$. The reason for this is the expansion of the universe; for example, the ratio of X bosons to photons $n_X/n_\gamma = \eta$ has a deviation Δ from its equilibrium value η_{eq} (of course at a given T) that satisfies a differential equation

$$\Delta' = \frac{d\Delta}{dz} = \frac{d}{dz}(\eta - \eta_{eq}) = -zKf(z)\Delta - \eta'_{eq}, \tag{29}$$

where $z = M_X/T$ and $f(z)$ is a positive function of z which parametrizes the decays and the scatterings. For $z \gtrsim 1$, $\eta_{eq}(z) \approx (\pi/2)^{1/2}z^{3/2}e^{-z}$; the expansion of the universe causes the temperature to drop and hence $\eta'_{eq} \neq 0$. Qualitatively, if η'_{eq} equaled zero and $K \gg 1$, the solution of (29) would be that Δ would relax exponentially to zero. As is, we have Δ' relaxing to zero and

$$\Delta \approx -\frac{\eta'_{eq}}{zKf(z)} \approx \frac{\eta_{eq}}{zKf(z)}. \tag{30}$$

A similar equation is set up for $B = (n_B - n_{\bar{B}})/n_\gamma$, and we find a nonzero B even for $K \gg 1$.

Q.16) I'm ready to quit. One final question, though. How did Sakharov happen to think of this problem and manage to identify the key ingredients in 1967, before one even had GUT theories, etc.?

A.16) Well, part of the answer is that he's very smart and very original, but it's not as if he pulled it completely out of thin air. Weinberg, for instance, in the concluding speculations in his 1964 Brandeis lectures [21] says that it seems plausible to him that the constitution of the universe is a consequence of the lack of conservation of baryon and lepton numbers.

Sakharov's article [22] is fascinating. Some parts of it are certainly different from what we presently believe, i.e., he postulates a combined baryon-muon charge $n = 3$(no. baryons) $-$ (no. muons). On the other hand he identifies the correct ingredients for baryon asymmetry, including following up an idea of Markov's of superheavy spinless particles in the theory, and concludes his paper by estimating the lifetime for proton decay $p \to \mu^+ \nu_\mu \nu_\mu$ via a diagram very similar to what we presently use; i.e., he has a fractionally charged gauge boson that connects quarks to leptons! He finds the lifetime of the proton to be 10^{50} years. He also speculates in this article that the greater part of the mass of the universe may be due to muon neutrinos and that therefore $m_{\nu_\mu} \approx 30$ ev! By the way in 1967, he also wrote another interesting note [23], called "Vacuum Quantum Fluctuations in Curved Space and the Theory of Gravitation," which bears a relation to recent work of Adler and Zee [24] in trying to derive field-theoretically Newton's constant G. Yet a third 1967 note [25] discusses $K^\circ \bar{K}^\circ$ transitions by what we would nowadays call a $2W$-exchange box diagram. Not bad!

References

[1] K. A. Olive, D. N. Schramm, G. Steigman, M. S. Turner, and J. Yang, *Astrophys. J.* 246: 557 (1981).

[2] See, e.g., S. Weinberg, *Gravitationand Cosmology*, Wiley, New York (1972), ch. 15.

[3] This scenario was proposed by S. Weinberg, *Phys. Rev. Lett.* 42: 850 (1979), and D. Toussaint, S. B. Treiman, F. Wilczek, and A. Zee, *Phys. Rev. D* 19: 1036 (1979). Earlier relevant work had been done by M. Yoshimura, *Phys. Rev. Lett.* 41: 281 (1978); A. Ignatiev, N. Krasnikov, V. Kuzmin, and A. Tavkhelidze, *Phys. Lett. B* 76: 486 (1978); S. Dimopoulos and L. Susskind, *Phys. Rev. D* 19 (1978); J. Ellis, M. K. Gaillard, and D. V. Nanopoulos, *Phys. Lett. B* 80: 360 (1979); and N. J. Papastamatiou and L. Parker, *Phys. Rev. D* 19: 2283 (1979); we will later refer to even earlier works.

[4] See, e.g., J. N. Fry, K. A. Olive, and M. S. Turner, *Phys. Rev. D* 22: 2953 (1980).

[5] G. Steigman, *Ann. Rev. Astron. Astrophys.* 14: 339 (1976).

[6] A. Buffington and S. M. Schindler, *Ap. J. Lett.* 247: 105 (1981).

[7] As examples of these discussions see, e.g.: A. H. Guth and E. Weinberg, *Phys. Rev. D* 23: 877 (1981); E. Witten, *Nucl. Phys. B* 177: 477 (1981); and P. Steinhardt, *Nucl. Phys. B* 179: 492 (1981).

[8] J. Ellis and G. Steigman, *Phys. Lett. B* 89: 186 (1980).

[9] E. W. Kolb and S. Wolfram, *Nucl. Phys. B* 172: 224 (1980); *Phys. Lett. B* 91: 217 (1980).

[10] A. D. Dolgov and Ya. B. Zeldovich, *Rev. Mod. Phys.* 53: 1 (1981).

[11] D. V. Nanopoulos and S. Weinberg, *Phys. Rev. D* 20: 2484 (1979).

[12] For a review see P. Langacker, *Phys. Rep.* 72: 187 (1981).

[13] S. Barr, G. Segrè, and H. A. Weldon, *Phys. Rev. D* 20: 2494 (1979). See also J. Ellis et al., ref. [3]; A. Yildiz and P. H. Cox, *Phys. Rev. D* 21: 906 (1980).

[14] A. Buras, J. Ellis, M. K. Gaillard, and D. V. Nanopoulos, *Nucl. Phys. B* 155: 189 (1979).

[15] G. Segrè and M. S. Turner, *Phys. Lett. B* 99: 399 (1981).

[16] R. Peccei and H. Quinn, *Phys. Rev. Lett.* 38: 1440 (1977); *Phys. Rev. D* 16: 1791 (1977).

[17] R. Slansky, in *First Workshop on Grand Unification*, edited by P. Frampton, S. Glashow, and A. Yildiz, Math-Sci Press, Brookline, Mass. (1980), p. 57.

[18] V. A. Kuzmin and M. E. Shaposhnikov, *Phys. Lett. B* 92: 115 (1980); T. Yanagida and M. Yoshimura, *Phys. Rev. D* 23: 2048 (1981); and H. Haber, G. Segrè, and S. Soni, *Phys. Rev. D* 25: 1400 (1982). See however, e.g., A. Masiero and R. N. Mohapatra, *Phys. Lett. B*, to be published, for a scheme where light N decays lead to a baryon asymmetry. For a detailed discussion of n/s in various models as well as many other topics see the excellent lecture notes on cosmological baryon production by M. Yoshimura in *Proceedings of the Fourth Kyoto Summer Institute: Grand Unified Theories and Related Topics*, edited by M. Konuma and T. Maskawa, World Scientific, Singapore (1981).

[19] A more general discussion of this point is given in A. Masiero, R. N. Mohapatra, and R. D. Peccei, MPI-PAE report 25/81. They emphasize the point that a phase transition necessarily takes place between the time of baryosynthesis and our present universe, namely the $SU_2 \times U_1$ one. This means that high-temperature and low-temperature CP-violating phases are not connected.

[20] See, e.g., M. S. Turner, 1981 Les Houches Summer School on Gauge Theories, to be published. For detailed calculations see, e.g., refs. [9] and [4]; J. N. Fry, K. A. Olive, and M. S. Turner, *Phys. Rev. D* 22: 2977 (1980); and J. A. Harvey, E. W. Kolb, D. B. Reiss, and S. Wolfram, L.A.U.R. 81-2336, to be published in *Nucl. Phys.*

[21] S. Weinberg in *Lecture on Particles and Fields*, edited by S. Deser and K. Ford, Prentice-Hall, Englewood Cliffs, N.J. (1964), p. 482.

[22] A. Sakharov, *JETP Lett.* 5: 24 (1967).

[23] A. Sakharov, *Sov. Phys. Dok.* 112: 1040 (1968).

[24] S. L. Adler, *Phys. Rev. Lett.* 44: 1567 (1980). A. Zee, *Phys. Rev. Lett.* 42: 417 (1979); *ibid.* 44: 703 (1980); *ibid.* 48: 295 (1982).

[25] A. Sakharov, *JETP Lett.* 5: 27 (1967).

Speculations on the Origin of the Matter, Energy, and Entropy of the Universe

I. Toward the Inflationary Universe

I first began to work for Francis as a graduate student during the academic year 1968–69. Francis had spent the previous year on leave at the Institute for Advanced Study, working with Murray Gell-Mann, Murph Goldberger, and Norman Kroll on the problems of the theory of weak interactions; in particular, they were concerned about the appearance in higher-order perturbation theory of large neutral strangeness changing currents. Although they failed to provide the right solutions to these problems, they were very accurate in identifying the important problems and in indicating the general nature of the solutions which were later found. In its assumptions and methods of analysis, their paper [1] foreshadowed the incredibly exciting decade of progress which would follow.

Gell-Mann et al. [1] calculated the lowest-order amplitude for the process $v\bar{v} \to W^+ W^-$ in the intermediate vector boson model, and determined that the partial-wave amplitude would exceed the unitarity bound when the center of mass energy exceeded $(24\pi/G_F)^{1/2}$. They concluded that the weak interactions would have to be modified on this energy scale in order to remain unitary. Furthermore, they determined that the cancellation of the neutral strangeness changing current would require a modification at a much lower energy scale—not too different from a few nucleon masses.

The necessary modification at the mass scale $G_F^{-1/2}$ was of course provided by the Weinberg-Salam theory. Gell-Mann et al. (along with most of the physics community) were apparently unaware of Weinberg's 1967 paper [2]. (In the Weinberg–Salam theory, the leading high-energy behavior of $v\bar{v} \to W^+ W^-$ is canceled by the diagram with a Z^0 intermediate state.) The low-mass modification was provided by the introduction of the charmed quark, as developed in 1970 by Glashow, Iliopoulos, and Maiani [3]. (I recall that Francis was very interested in this paper when it first appeared. It might have become the subject of my thesis, if I could have only thought of something to do with it.)

During the first half of the 1970s, progress in the field of particle theory was astounding. QCD became the accepted model of the strong interactions, and the work of Weinberg, Salam, and Glashow was synthesized into the accepted model of the weak and electromagnetic interactions. While only one of the four basic interactions of nature was thought to be understood when I was a graduate student, we now had an acceptable theory for three of the four. It was from this atmosphere of enormous success that the chutzpah necessary for grand unified theories (GUTs) [4] emerged.

Before grand unification, our understanding of particle physics consisted of the $SU_3 \times SU_2 \times U_1$ model of strong, weak, and electromagnetic interactions, combined with classical general relativity as a description of gravity. When these theories are extrapolated to high energies, the first inconsistency that appears to occur is the breakdown of classical gravity at the order of the Planck mass,[1] $M_P \equiv G^{-1/2} = 1.22 \times 10^{19}$ GeV. This is quite analogous to the breakdown of the old weak interaction theory at energies of the order of $G_F^{-1/2}$. Just as Gell-Mann et al. assumed that the flaws of the weak-interaction theory would be evident in its high-energy extrapolation, it is tempting to assume that the only flaws in our current theories lie near the Planck mass. Since the three gauge-coupling constants for the SU_3, SU_2, and U_1 interactions appear to be about equal [5, 6] at an energy scale of about 4×10^{14} GeV, it is also tempting to assume that these interactions are unified in a single gauge group G, which is spontaneously broken so that the gauge particles which lie outside the $SU_3 \times SU_2 \times U_1$ subgroup acquire masses $M_X \approx 4 \times 10^{14}$ GeV. The first model of this type was the SU_5 model of Georgi and Glashow [7], proposed in 1974. Quantitatively, the assumption that all three coupling constants have equal values at some energy results in a prediction [8] of $\sin^2 \theta_W$ (where θ_W is the Weinberg angle) of 0.209 ± 0.005. The experimental value [8] is 0.215 ± 0.012, so there is a strong hint that grand unification is on the right track.

---

1. I use units for which $\hbar = c = k$ (Boltzmann constant) $= 1$. Then 1 m $= 5.068 \times 10^{15}$ GeV^{-1}, 1 kg $= 5.610 \times 10^{26}$ GeV, 1 sec $= 1.519 \times 10^{24}$ GeV^{-1}, and $1°$K $= 8.617 \times 10^{-14}$ GeV.

If grand unification is accepted as being at least plausible, then one is lead very quickly to the rather far-reaching questions which are raised in my title. First, there is the question of the baryon number of the universe. If baryon number were exactly conserved, then the baryon number of the observed universe (estimated to be $10^{79\pm1}$) would have been completely fixed by the initial conditions. However, it was noticed very early that the absence of baryon number conservation in GUTs would mean that the ratio of baryon number to entropy (n_B/s) in the universe (which is about $10^{-10\pm1}$) would be determined [9] by particle interactions in the very early universe ($t \approx 10^{-35}$ sec) and would be essentially independent of the initial conditions. There is of course no precise prediction from GUTs; we are talking about a class of theories, and each theory contains a number of undetermined parameters. The theoretical predictions for n_B/s range from [10] about 10^{-21} to [11] 10^{-4}, so the desired value is well within the predicted range. Obviously one cannot claim a quantitative success; but one can claim a plausible solution to a question which profoundly affects the nature of our universe.

However, the nonconservation of baryon number has implications which transcend the statement that the ratio n_B/s is in principle calculable. The important point is that baryon number is the only possibly conserved quantity for which the universe appears to have a nonzero value. Thus, in the absence of baryon-number conservation, it becomes possible that our universe emerged from practically nothing at all [12].

The above statement may be surprising to many, since there is a common misconception that the universe has a large conserved energy. However, the global conservation of energy in the context of general relativity is a rather subtle concept. In a spacetime which is asymptotically Minkowskian, it is possible to define a conserved energy and to show (under plausible assumptions) that it is positive semidefinite, and vanishes only if the space is empty and Minkowskian [13]. Our universe, however, is not believed to be asymptotically Minkowskian. If the universe is described by a Robertson-Walker metric, as is generally assumed in simple models, then there is no globally conserved energy.

In the inflationary universe scenario [14], which will be described in section III, the observed universe can evolve from a region of space which is so small that it would have contained no particles at all. In this sense, one would have an explanation for the origin of all the matter, energy, and entropy in the observed universe.

II. The Monopole, Flatness, and Horizon Problems of the Standard Model

There is a rather well-defined standard model of the early universe; it is an adiabatically expanding radiation-dominated universe described by a Roberston-Walker metric. In this section I will review the basics of this model, emphasizing three of its difficulties: the monopole, flatness, and horizon problems. The monopole problem is a dynamical problem associated with GUTs. Using a typical GUT and the standard cosmological model, one estimates that far too many magnetic monopoles would have been produced. The flatness and horizon problems are associated with the initial conditions of the universe. That is, the initial conditions which are required by the standard model have to be fixed in a highly precise manner. It seems (at least to me) to be very difficult to accept these conditions without an explanation of the mechanism which created them.

The assumption that the universe is homogeneous and isotropic implies that it can be described in comoving coordinates by the Robertson-Walker metric:

$$ds^2 = -dt^2 + R^2(t)\left[\frac{dr^2}{1 - kr^2} + r^2(d\theta^2 + \sin^2\theta d\phi^2)\right]. \qquad (2.1)$$

The universe is closed if $k > 0$, open if $k < 0$, and flat if $k = 0$. I will follow the custom of rescaling r and $R(t)$ so that k takes on only one of the discrete values $+1$, -1, or 0. The evolution of $R(t)$ is governed by the Einstein field equations, which reduce in this case to

$$\ddot{R} = -\frac{4\pi}{3}G(\rho + 3p)R, \qquad (2.2a)$$

$$\left(\frac{\dot{R}}{R}\right)^2 = \frac{8\pi}{3}G\rho - \frac{k}{R^2}, \qquad (2.2b)$$

where ρ is the energy density, p is the pressure, and the dot denotes differentiation with respect to t. The quantity $H \equiv \dot{R}/R$ is known as the Hubble "constant." Its value today is denoted by

$$H_0 = h_0 \times (100\text{km/sec-Mpc})$$
$$= h_0 \times (9.78 \times 10^9 \text{ yr})^{-1}, \qquad (2.3)$$

where empirically [15] h_0 probably lies between 0.4 and 1. (I will use the subscript 0 to denote the present value of any given quantity.) Equation (2.2b) allows one to define a critical density

$$\rho_c = 3H^2/8\pi G, \tag{2.4}$$

which is the energy density which gives a flat ($k = 0$) universe. Its value today is $\rho_{c0} = 1.87 \times 10^{-29} \, h_0^2$ gm/cm^3. The ratio ρ/ρ_c is denoted by Ω, and its value today is believed to lie in the range [16]

$$0.1 \lesssim \Omega_0 \lesssim 1. \tag{2.5}$$

Conservation of energy is expressed by

$$\frac{d}{dt}(R^3\rho) = -p\frac{d}{dt}(R^3), \tag{2.6}$$

and the assumption of adiabaticity (conservation of entropy) takes the form

$$\dot{S} = 0, \tag{2.7a}$$

where

$$S \equiv R^3 s, \tag{2.7b}$$

and s denotes the entropy density; S will be called the characteristic entropy.

To determine the evolution of the universe, the above equations must be supplemented by an equation of state for matter. The assumption of thermal equilibrium implies that matter can be described by a free-energy density $\mathcal{F}(T, \mu_i)$, where T denotes the temperature, and each μ_i denotes the chemical potential associated with some conserved quantity Q_i. One then has

$$
\begin{aligned}
s &= -\partial\mathcal{F}/\partial T, \\
q_i &= -\partial\mathcal{F}/\partial\mu_i, \\
\rho &= \mathcal{F} + Ts + \sum_i \mu_i q_i, \\
p &= -\mathcal{F},
\end{aligned}
\tag{2.8}
$$

where q_i denotes the density of the conserved quantity Q_i. In the standard model one assumes that the very early universe is dominated by an ideal quantum gas of essentially massless particles, with $\mu_i = 0$ for all i. In that case

$$\mathcal{F}(T) = -\frac{\pi^2}{90}\left[N_b(T) + \frac{7}{8}N_f(T)\right]T^4, \tag{2.9}$$

where $N_b(T)$ denotes the number of bosonic spin degrees of freedom which are effectively massless at temperature T; each possible helicity contributes one unit, and particle and antiparticle count separately if they are distinguishable; $N_f(T)$ denotes the corresponding number for fermions.

One is also sometimes interested in particle number densities, given by

$$n = \frac{\zeta(3)}{\pi^2}\left(N_b(T) + \frac{3}{4}N_f(T)\right)T^3, \tag{2.10}$$

where $\zeta(3) = 1.20206\ldots$ is the Riemann zeta function. An important quantity which helps to characterize our universe is the ratio of the baryon number density n_B to the photon density n_γ. Taking the current photon temperature as

$$T_{\gamma 0} = \tau_0 \times 2.7°\text{K} \tag{2.11}$$

(where empirically τ_0 probably lies in the range 1.0–1.1), one has

$$\begin{aligned}\frac{n_B}{n_\gamma}\bigg|_0 &= \frac{3\pi}{16\zeta(3)}\frac{\Omega_{B0}H_0^2 M_P^2}{m_B T_{\gamma 0}^3} \\ &= 2.81 \times 10^{-8}h_0^2\Omega_{B0}\tau_0^{-3},\end{aligned} \tag{2.12}$$

where $m_B = 938$ GeV is the mass of a baryon, $M_P \equiv G^{-\frac{1}{2}} = 1.22 \times 10^{19}$ GeV is the Planck mass, and Ω_{B0} is the current ratio of the baryonic mass density to the critical mass density. Arguments based on the abundances of helium and deuterium produced in the big bang suggest that [17] $(n_B/n_\gamma)_0 \lesssim 4.2 \times 10^{-10}$, so $\Omega_{B0} \lesssim 0.014h_0^{-2}\tau_0^3$.

One can now summarize the history of the universe. At very early times, the curvature term $-k/R^2$ on the right-hand side of (2.2b) is negligible. (This point will be examined quantitatively later in this section). Matter will be described by the free-energy density (2.9); for $T \gtrsim 10^{14}$ GeV, all of the degrees of freedom of the GUT are effectively massless. As long as $N_b(T)$ and $N_f(T)$ are treated as constants, adiabaticity implies that RT is a constant. Equation (2.2b) can then be rewritten as

$$\dot{T} = -\gamma T^3/M_P, \tag{2.13}$$

where

$$\gamma^2 = \frac{4\pi^3}{45}\left(N_b + \frac{7}{8}N_f\right). \tag{2.14}$$

For the minimal SU_5 model [7] with three generations of fermions, $N_b = 82$, $N_f = 90$, and $\gamma \approx 21$. The solution is then

$$T^2 = M_P/(2\gamma t). \tag{2.15}$$

Since RT is a constant, it follows that $R(t) \propto t^{1/2}$. When T is 10^{14} GeV, these equations give $t \approx 1.9 \times 10^{-35}$ sec, and $\rho \approx 1.2 \times 10^{75}$ gm/cm^3.

As the universe cools through the temperature range $T \approx 10^{14}$ GeV, grand unified theories predict the existence of a phase transition [18] (or perhaps several phase transitions). It is at this point that the Higgs field acquires its nonzero expectation value. In the standard model it is assumed that this phase transition occurs quickly, with only a negligible amount of supercooling. In that case, causality considerations alone imply that it is impossible for the Higgs field to align uniformly over large distances. If one sets $k = 0$ in the Robertson-Walker metric (2.1), one finds that light pulses (with $ds^2 = 0$) travel a coordinate distance $r(t_2, t_1)$ between time t_1 and time t_2, with

$$r(t_2, t_1) = \int_{t_1}^{t_2} \frac{dt}{R(t)}. \tag{2.16}$$

The quantity

$$r_H(t) \equiv r(t, 0) \tag{2.17}$$

is the coordinate horizon length; two objects which are separated by more than this coordinate distance have had no chance to communicate since the origin of the universe. The physical horizon length is given by

$$d_H(t) = R(t)r_H(t). \tag{2.18}$$

Using $R(t) \propto t^{1/2}$ for the early universe, it follows that $d_H = 2t$. Thus the Higgs-field correlation length ξ cannot be longer than $2t$.

The absence of correlation in the Higgs field is associated with the presence of magnetic monopoles. These monopoles are characterized by topologically stable knots in the Higgs expectation value. Thus, one expects to find (very roughly) one monopole in each cube of volume ξ^3. Since $\xi < 2t$, the density n_M of monopoles is greater than $(2t)^{-3}$. Using (2.15), (2.8), and (2.9), this inequality can be rewritten as

$$n_M/s \gtrsim 10^2 (T/M_P)^3, \tag{2.19}$$

where I have estimated N_b and N_f from the SU_5 GUT. For $T \approx 10^{14}$ GeV, this gives $n_M/s \gtrsim 10^{-13}$. Preskill [19] has shown that such a density of monopoles would not be noticeably diminished by annihilation processes. If they were formed originally, they would have survived to the present day. Since the mass of a monopole is about 10^{16} GeV, this ratio of n_M/s

would give a mass density to the present universe which is about 10^{12} times the critical mass density. This is the monopole problem.

I should point out that Langacker and Pi [20] have shown that the monopole problem can be avoided by introducing a more complicated Higgs structure with parameters adjusted so that the U_1 symmetry of electromagnetism is broken at high temperatures (i.e., a high-temperature superconductor). This model, however, is somewhat complicated and unattractive.

For temperatures above 10^{14} GeV, baryon-number-violating processes are rapid. This baryon number density quickly approaches its thermal equilibrium value, which is guaranteed by TCP symmetry to vanish identically. The creation of a net baryon number density requires a departure from thermal equilibrium. It is usually assumed [9] that this departure from thermal equilibrium is provided by the number density of some particle which I will denote by X—it is usually taken to be a color-triplet Higgs particle. When T falls below M_X, the number density n_X would decline sharply in a thermal-equilibrium mixture. However, if the decay rate is slow or comparable to the age of the universe at this time, then n_X will be larger than its thermal-equilibrium value. When the X particle finally decays, the CP violations in its couplings to fermions provide a tendency to produce an excess of quarks over antiquarks. The excess tends to be small because it vanishes at tree level.

The universe continues to expand and cool, and more degrees of freedom freeze out. When the temperature reaches 1 MeV, the effectively massless degrees of freedom are the photons, electrons, positrons, and neutrinos. If there are three species of neutrinos, then $\gamma = 5.44$ and $t = 0.74$ sec. Estimates of the neutrino interaction rates indicate that it is at about this time that the neutrinos decouple (i.e., lose thermal contact) with the rest of matter. From this time on, the neutrinos can be treated as a collisionless gas. It follows that the neutrinos continue to be described by a thermal distribution, with RT_ν maintaining the same value it had before decoupling. When $T \approx \frac{1}{2}$ MeV (at $t \approx 3.0$ sec), the electron-positron pairs disappear quickly from the thermal equilibrium mixture. Their entropy (with $s_{e^+ e^-} \propto \frac{7}{8} N_f T^3 = \frac{7}{2} T^3$) is imparted to the photons (with $s_\gamma \propto N_b T^3 = 2T^3$), and as a result RT_γ is increased by a factor of $\left(\frac{11}{4}\right)^{1/3}$.

The ratio

$$T_\gamma / T_\nu = (11/4)^{1/3} \approx 1.40 \tag{2.20}$$

is then preserved until the present. The ratio of total entropy (photons and neutrinos) to photon number is then fixed at $s/n_\gamma = 7.04$. Equation (2.15) will now hold with $T = T_\gamma$ and $\gamma = 3.04$. (To be completely precise, the quantity t would have to be replaced by $t + \delta t$, where δt is of the order of 3 sec. The value of δt could be found by numerically integrating (2.2b) during the period of electron-positron pair annihilation).

I am now ready to discuss the flatness problem, which is associated with the value of the characteristic entropy S defined by (2.7b). Since S is conserved, its value today can in principle be determined by current observations. However, the value of R_0 (the radius of curvature of the present universe) is highly uncertain. By (2.2b) and (2.4) it can be written as

$$R_0 = \frac{1}{H_0 \sqrt{|1 - \Omega_0|}}. \tag{2.21}$$

Since Ω_0 could be arbitrarily close to one, R_0 could be arbitrarily large. If we assume that $\Omega_0 < 2$, then we can conclude that

$$R_0 > H_0^{-1} \approx 10^{10}\,\text{yr}. \tag{2.22}$$

Using this bound and the entropy density s of photons and neutrinos, one finds

$$S > 2 \times 10^{87}. \tag{2.23}$$

The flatness problem is essentially the problem of understanding why S is such a large number. Since it is a dimensionless constant fixed by the initial conditions of the model, it seems more reasonable to expect it to be of order unity. Furthermore, the large value of S is associated with extreme fine tuning of the Hubble expansion rate to the mass density in the early universe. To see this, note that (2.2b) can be rewritten to give

$$\frac{\Omega - 1}{\Omega} = \frac{3}{8\pi G} \frac{ks^{2/3}}{\rho S^{2/3}}. \tag{2.24}$$

Numerically, one finds that $\Omega = 1 \pm O(10^{-15})$ when $T = 1$ MeV, and $\Omega = 1 \pm O(10^{-49})$ when $T = 10^{14}$ GeV. It is certainly consistent with the spirit of modern physics to assume that any equality of this precision requires an explanation.

To discuss the horizon problem it is necessary to carry the history of the universe through to the present time. For simplicity, I will now assume that the universe contains a critical density of massive particles. (More general results are derived in the appendix.) The mass density of these particles as a function of temperature is then

$$\rho_m = \rho_{c0} \left(\frac{T_\gamma}{T_{\gamma 0}} \right)^3. \tag{2.25}$$

This term will equal the energy density ρ_r of radiation of massless particles when $T_\gamma = 5.79\, h_0{}^2 \tau_0{}^{-3}$ eV, which according to the evolution equation occurs when $t = 1.25 \times 10^3\, \tau_0{}^6\, h_0{}^{-4}$ yr. From that point on the universe would be matter-dominated, in which case (2.2b) implies that $R(t) \propto t^{2/3}$. Then $H = \dot{R}/R = 2/(3t)$, so the present age of the universe is

$$t_0 = \frac{2}{3} H_0{}^{-1} = 6.52 \times 10^9 h_0{}^{-1} \text{ yr}. \tag{2.26}$$

(Studies [21] of ^{232}Th/^{238}U and ^{187}Re/^{187}Os abundance ratios imply that $t_0 > 8.7 \times 10^9$ yr. Thus, if $\Omega_0 = 1$, then $h_0 < 0.75$.) The time-temperature relation during this matter-dominated era is then given by

$$T_\gamma^3 = \frac{4 T_{\gamma 0}{}^3}{9 H_0{}^2 t^2}. \tag{2.27}$$

When the temperature falls to about 4000°K (0.34 eV), the electrons combine with the protons to form neutral hydrogen [22]. This process is known as recombination, although the prefix "re-" seems totally inappropriate. According to (2.27), recombination occurs at $t = 1.14 \times 10^5$ $\tau_0 h_0{}^{-1}$ yr. Thus, if the universe is closed, then it was matter-dominated well before the time of recombination.

At recombination there is a sharp drop in the opacity of matter. The mean free time for a photon is then much longer than the age of the universe, and from this point on the photons of the microwave background radiation can be treated as a collisionless gas. Thus, the extreme isotropy which is observed in the microwave background radiation today is believed to have been present at the time of recombination.

The horizon problem is the statement that the isotropy of the cosmic microwave background radiation is an apparent violation of causality. To see this, let us suppose that the integral in (2.16) for $r(t_2, t_1)$ is dominated by the contribution from the matter-dominated era. Then the coordinate horizon length $r_H(t) \propto t^{1/3}$, and the physical horizon length $d_H(t) = 3t$. The cosmic background radiation can be thought of as having been emitted at the time of recombination t_r. Thus, the source of this radiation is separated from us by a coordinate distance $r(t_0, t_r) = r_H(t_0) - r_H(t_r)$. Now consider two microwave antennas pointed in opposite directions. The sources of the photons arriving at the antennas were separated from each other at the time of emission by N horizon lengths, where

$$N = 2\left[\frac{r_H(t_0) - r_H(t_r)}{r_H(t_r)}\right] \qquad\qquad (2.27a)$$

$$= 2\left[\left(\frac{T_{\gamma r}}{T_{\gamma 0}}\right)^{1/2} - 1\right] \approx 75. \qquad\qquad (2.27b)$$

The problem is to explain how two regions of space, separated by 75 horizon lengths, came to be at the same temperature at the same time. In the context of the standard model, this long-range homogeneity has to be simply assumed as part of the initial conditions.

In the above discussion I assumed for simplicity that $\Omega_0 = 1$ and that the universe was matter-dominated throughout the relevant portion of its history. Since these assumptions are highly restrictive, the result may not be totally convincing. I have therefore repeated these calculations in the appendix without the use of either simplifying assumption. It is found that for reasonable values of the parameters, $N > 90$.

Throughout my discussion (with the exception of the appendix) I have neglected the curvature term k/R^2 in (2.2b). One can easily show that this term will exceed the contribution ρ_m due to massive particles when

$$T_\gamma < \frac{|1 - \Omega_0|}{\Omega_0} T_{\gamma 0}. \qquad\qquad (2.28)$$

Even if Ω_0 is as small as 0.01, the curvature term remains unimportant until T_γ falls to the order of 100 $T_{\gamma 0}$.

The horizon and flatness problems are both statements about the initial conditions required by the standard model. It is therefore possible that the solution to these problems lies in the era of quantum gravity, at times so early that the assumptions of the standard model are not yet applicable. However, it will be shown in the next section that these problems might be solved at the level of grand unified interactions, using the inflationary universe scenario. Even if the the solution really occurs at the level of quantum gravity, it is still conceivable that the mechanism closely resembles the inflationary scenario.

III. The Inflationary Universe

The inflationary scenario is an attempt to avoid the three problems discussed in the previous section. For the scenario to work, the underlying particle theory must have certain features. First, the theory must have a high-temperature phase which can supercool into a long-lived metastable state with an energy density $\rho_v > 0$. The inflation will take place while the observable part of the universe is trapped in this metastable state. Second, the metastable state must have a way of smoothly decaying, releasing the energy density ρ_0 as thermal radiation. This decay, however, must not occur too quickly. Finally, the theory must have a mechanism for producing an acceptable baryon-number to entropy ratio at temperatures which are achieved *after* the decay of the metastable state.

Standard GUTs typically satisfy the first and third of the above conditions. The metastable state is simply the false vacuum which arises when the Higgs fields settle into a local minimum of the potential. This state typically has an energy density ρ_0 of order $(10^{14} \text{ GeV})^4$. If this energy is thermalized it will produce a temperature of order 10^{14} GeV, and the standard scenario of baryon production can follow.

The second point, the smooth decay of the false vacuum, is still a problematical requirement. I will return to this question later.

Let me first describe the inflationary scenario [14]. For simplicity, I will begin by assuming that the universe is initially described by a Robertson-Walker metric, and that the matter content is initially in thermal equilibrium at high temperatures. I will argue later that these assumptions can probably be discarded. I perfer to avoid the uncertainties of Planck-mass physics, so I will begin the description at a temperature T_i which might be of the order of 10^{17} GeV. The universe will expand and cool, with T falling to the critical temperature T_c and then supercooling below it. The energy and entropy densities are well approximated during this period by

$$\rho(T) = cT^4 + \rho_v \tag{3.1}$$

and

$$s(T) = \tfrac{4}{3}cT^3, \tag{3.2}$$

where

$$c = \frac{\pi^2}{30}\left(N_b + \frac{7}{8}N_f\right) \tag{3.3}$$

The term cT^4 represents the radiation of massless particles: for the minimal SU_5 model [7], $c \approx 53$. The quantity ρ_v is the energy density of the metastable state, and is typically [23] of order T_c^4. Thus the ρ_v term is important only when supercooling takes place. When the ρ_v term dominates the right-hand side of (2.2b), the solution has the simple form

$$R(t) \propto e^{\chi t}, \tag{3.4}$$

where

$$\chi = \left(\frac{8\pi\rho_v}{3M_P^2}\right)^{1/2} \approx \frac{T_c^2}{M_P}. \tag{3.5}$$

(This is the metric of a de Sitter space [24].) It is also possible to solve (2.2b) exactly, using the energy density (3.1). The solutions are characterized by k and the characteristic entropy S:

$$R^2(t) = A \sinh \chi t (\cosh \chi t - \lambda \sinh \chi t) \tag{3.6}$$

where

$$A = 2\sqrt{\frac{c}{\rho_v}}\left(\frac{3S}{4c}\right)^{2/3} \tag{3.7}$$

and

$$\lambda = \frac{3M_P^2}{16\pi\sqrt{c\rho_v}}\left(\frac{4c}{3S}\right)^{2/3} k. \tag{3.8}$$

If $\lambda < 1$, these solutions all approach the form (3.4) for large time. If $\lambda > 1$, the universe would contract to a singularity in a finite time. In that case, there is a second solution in which the universe contracts from infinite size and then bounces, given by

$$R^2(t) = \tfrac{1}{2}A(\lambda + \sqrt{\lambda^2 - 1}\cosh 2\chi t). \tag{3.9}$$

(This solution, however, does not appear to me to be a plausible beginning for our universe.) My prejudice is that the initial value of S should be of order unity, which implies that $|\lambda|$ is of order $(M_P/T_c)^2 \approx 10^{10}$. Thus a long-lived universe would require $k = -1$.

Suppose now that the universe reaches the critical temperature and then expands by a factor of Z before undergoing the phase transition. I will temporarily assume that when the phase transition finally occurs, it happens quickly and leads to the thermalization of the energy density ρ_v.

When the energy density is thermalized, the temperature rises to a value on the order of T_c. The entropy density s is then comparable to its value when the universe first cooled to T_c, but $R(t)$ has increased since this time by a factor Z. The characteristic entropy S is therefore increased by a factor of roughly Z^3. If $Z > 10^{29}$, then the fact that $S > 10^{87}$ becomes natural; the flatness problem is solved.

I want to emphasize that values of $Z \gg 10^{29}$ are also quite acceptable; it would imply that $S \gg 10^{87}$ (i.e., that the universe is very close to flat).

In fact, it seems very likely that any mechanism which can explain why $S > 10^{87}$ will probably overshoot by many orders of magnitude.

The horizon problem also disappears, because in this scenario the region of space in the early universe which evolves to become our observed universe was smaller than in the standard model by a factor of Z. If Z is sufficiently large, then the observed universe can come into causal contact at the beginning of the period of exponential expansion. Quantitatively, suppose the era of exponential expansion extends from time $t_c\text{-}\Delta t$ to time t_c, where $Z = \exp(\chi\Delta t)$. During this time $R(t) = R_c \exp(\chi t)$, and the contribution from this era to $r_H(t_c)$ is given by $(Z - 1)\exp(-\chi t_c)/(\chi R_c)$. Then the physical horizon distance

$$d_H(t_c) \gtrsim Z\chi^{-1}. \tag{3.10}$$

The total amount of entropy in the causally connected volume is then approximately $cd_H^3(t_c)T_c^3 \approx cZ^3 \ (M_P/T_c)^3 \approx 10^{17}Z^3$. If $Z > 10^{24}$, then this number would exceed the total entropy in the observed universe today. Since entropy is essentially conserved in this model from time t_c to the present, it follows that the observed part of the universe was causally connected at early times.

The monopole problem is somewhat more subtle, since it involves the details of the phase transition. In any case, the simple horizon argument which implied an excess of monopoles is evaded. I will return to this question.

There are two points which I now want to discuss in more detail: the assumed initial conditions, and the nature of the phase transition.

In my previous discussion, I assumed for simplicity that the universe began in a state which was already homogeneous and isotropic. If this assumption were really necessary, then it would be senseless to talk about "solving" the horizon problem; perfect homogeneity was assumed at the outset. The important point is that for our purposes the homogeneity can be very local. If the expansion factor $Z \approx 10^{29}$, then the Robertson-Walker metric need only be a good approximation to the true metric over a region which contains on the order of one particle. Since Z can be much larger than 10^{29}, the initial region may be so small that the expected number of particles it contains is far less than one. Thus, the initial universe can be quite chaotic, provided only that very small regions are locally homogeneous. The universe today would then also be very inhomogeneous when looked at in its totality, but the huge exponential expansion factor has transported the gross inhomogeneities to distance scales much greater than 10^{10} light-years.

As for anisotropy, this problem has been looked at by Barrow and Turner [25]. They find that anisotropies are strongly damped during the era of exponential expansion. Thus it is not really necessary to start with an isotropic metric, even within the small local region. (However, if the anistropy is large, it is possible that a larger value of Z may be required to smooth it out.)

Steigman [26] has pointed out that if the universe begins with a characteristic entropy S of order one, as envisioned in the inflationary scenario, then the calculation [27] of the time at which thermal equilibrium can be attained requires modification. Under these new circumstances, it turns out that particle collision rates are always slow compared to H^{-1}, and thermalization will never take place. While this observation certainly complicates the question of specifying the class of initial conditions which would lead to the inflationary scenario, I do not believe that it means that the class is small. Recall that the equilibrium state of matter in the exponentially expanding universe is no matter at all; and this state is achieved by the dilution of matter as the scale factor grows—a process which has nothing to do with collision rates. Of course, the inflationary scenario does require that the Higgs field gets trapped in the false-vacuum minimum of the potential, at least in some regions of space. But as the system loses energy due to the expansion of the universe, the Higgs field must tend to either the true vacuum or to a false vacuum. In thermal equilibrium (for some values of the Higgs potential parameters), the Higgs field would undoubtedly tend toward the false vacuum value [28]. In the absence of thermal equilibrium, I would still think this endpoint is at least possible, if not highly probable.

Finally, the time has come to discuss the nature of the phase transition itself. The simplest mechanism for the phase transition would be the Coleman-Callan [29] "fate of the false vacuum" process by which bubbles of the new phase are nucleated spontaneously by quantum tunneling at zero temperature. In my original inflationary universe paper [14], I pointed out that this mechanism would lead to gross inhomogeneities due to the randomness of the bubble-formation process. (I have since studied this question in more detail with Erick Weinberg [30], confirming the original conclusions. The situation is actually quite peculiar: the bubbles fill a fraction of the space which approaches one, but they remain always in finite-sized clusters, each of which is dominated by its largest bubble.) In the same paper I also relayed a suggestion given to me by Ed Witten: If the Higgs field potential is of the Coleman-Weinberg

[31] type, then it is conceivable that the phase transition might be driven after significant supercooling has taken place.

The dynamics of a phase transition of this type was subsequently studied by several authors [32, 33]. For the SU_5 model, the final word (see Sher's paper) appears to be that the bubble nucleation rate for the phase transition remains small for all temperatures for which it can be calculated. The calculations break down [33] when $T \approx 10^{7-8}$ GeV, at which point the saddle-point approximation fails.

Recently Linde [34] and Albrecht and Steinhardt [33] have suggested a modified ending for the inflationary scenario that I find very promising. These authors assume a Coleman-Weinberg potential for the Higgs field. Linde assumes that bubble formation takes place at a temperature of $\approx 2 \times 10^6$ GeV, determined by the scale at which the coupling constant for the unbroken SU_5 group becomes strong. Albrecht and Steinhardt assume that the phase transition takes place at $T \approx 10^{7-8}$ GeV, driven by many types of fluctuations which become significant when the saddle-point approximation fails. In both cases, the initial magnitude of the fluctuation in the Higgs field ϕ is of order T, which is significantly smaller than the expansion rate $\chi \approx 10^{10}$ GeV. The Higgs field is then assumed to follow its classical equations of motion, which means that it begins to "fall" very slowly toward the minimum of the potential. Meanwhile, however, the space continues to expand exponentially, with a time constand much faster than the time required for the Higgs field to "fall." The important result is that the single bubble or fluctuation can undergo the inflationary scenario; it can grow to a size much larger than the observed universe, and thus the entire observed universe is contained within one such region. Since the Higgs field is correlated throughout this region, the monopole problem is solved. Even when a discrete symmetry is spontaneously broken, there is no problem with domain walls. The horizon and flatness problems disappear as a consequence of the inflation.

One can easily imagine minor modifications of the above suggestion. For example, it is conceivable that temperature has nothing to do with it. Perhaps the symmetric phase remains metastable at zero temperature, with bubbles nucleating at a constant and slow rate. The magnitude of the initial fluctuations might be small; perhaps it is set by the scale at which the unbroken SU_5 gauge coupling becomes strong.

The details of the Linde-Albrecht-Steinhardt suggestion remain to be worked out. Most importantly, all the calculations that I have referred to have neglected the effect of gravitation on the process of bubble formation. This effect is likely to be very important. The de Sitter space has a horizon distance χ^{-1} (i.e., two comoving observers which are separated by a physical distance χ^{-1} will never be able to communicate), and the physical size of the bubbles is of order $T^{-1} \gg \chi^{-1}$. Furthermore, quantum fluctuations in the de Sitter space mimic thermal effects at a Hawking temperature [35] $\chi/2\pi$, and thus it is not clear that it makes sense to talk about temperatures smaller than this. A step toward taking into account these gravitational effects has been taken by Vilenkin [36] and by Hawking and Moss [41]. (Press [37] has also looked at the effects of Hawking radiation, and has developed a possibly workable scenario. However, there is some controversy concerning the properties of Hawking radiation that he assumed.)

To pursue the zero-temperature approach mentioned above, the right starting point is presumably the formalism of Coleman and De Luccia [38]. This leads to a remarkably elegant picture, since the interior of a bubble is described by an open Robertson-Walker metric. The formalism must be extended, however, to incorporate the higher-order corrections which are essential to the Coleman-Weinberg potential. Furthermore, it seems [39] that the action associated with any bubble solution is much less than one, indicating a breakdown of the saddle-point approximation.

IV. Conclusion

If the details of the Linde-Albrecht-Steinhardt suggestion work out, it would provide an elegant solution to some of the deepest problems in cosmology: the horizon problem, the flatness problem, the anistropy problem, and the magnetic-monopole problem. Such a scenario would also explain the origin of all the matter, energy, and entropy in the universe.

Probably the most striking recent development in the study of cosmology is the realization that the universe may be completely devoid of all conserved quantum numbers. If so, then even if we do not understand the precise scenario, it becomes very plausible that our observed universe emerged from nothing or from almost nothing. I have often heard it said that there is no such thing as a free lunch. It now appears possible that the universe is a free lunch.

Appendix. The Evolution of the Universe from $t = 3$ sec **Onward**

As discussed in section II, electron-positron pairs will disappear quickly when the temperature falls below $\frac{1}{2}$ MeV, an event which occurs roughly three seconds after the big bang. From this time onward the mass density of the universe has a very simple description. First, there is a fixed number of massless species. Taking three species of massless neutrinos and using the temperature relationship (2.20) gives

$$\rho_r = cT_\gamma^4, \tag{A.1}$$

where $c = 1.106$. Second, there is a density of massive particles that is proportional to R^{-3}, and can therefore be written as

$$\rho_m = \rho_{m0}\left(\frac{R_0}{R}\right)^3, \tag{A.2}$$

where ρ_{m0} is the present density of massive particles, and R_0 is the present value of the scale factor. In this appendix I will solve the evolution equations with these two contributions to the mass density, for curved as well as flat universes. My main interest will be in the horizon problem, which is intimately entwined with the evolution of the universe during this period. Since the simplifying approximations which were made in section III were in fact rather crude, I want to show the reader that the problem does not disappear with more precise calculations. In fact, it becomes slightly more severe.

A neutrino mass of order 50 eV would lead to only minor changes in this picture. Using (2.15) with $\gamma = 3.04$, one finds that the temperature would fall to the neutrino mass at

$$t = \left(\frac{50 \text{ eV}}{m_\nu}\right)^2 16.7 \text{ yr.} \tag{A.3}$$

From this time onward the evolution would obey the same description as above, except that the value of c would be reduced. If all three neutrino species were massive, the value of c would be reduced to its photon contribution of 0.658. The neutrinos would then of course contribute to the value of ρ_{m0}, but the value of this quantity is highly uncertain in any case.

Let Ω_m denote the current fraction of the critical mass density due to massive particles (deleting the subscript 0 used earlier), and let Ω_r denote the corresponding fraction due to the radiation of massless particles. The quantity Ω_r is then given by

$$\Omega_r = \frac{8\pi}{3}GcT_{\gamma 0}^4/H_0^2$$
$$= 4.011 \times 10^{-5}\tau_0^4 h_0^{-2}. \tag{A.4}$$

Defining Ω_c (c for curvature) by

$$\Omega_c = 1 - \Omega_m - \Omega_r, \tag{A.5}$$

(2.2b) can be rewritten as

$$\left(\frac{\dot{R}}{R}\right)^2 = H_0^2\left[\Omega_r\left(\frac{R_0}{R}\right)^4 + \Omega_m\left(\frac{R_0}{R}\right)^3 + \Omega_c\left(\frac{R_0}{R}\right)^2\right]. \tag{A.6}$$

I will not need the explicit solution to this equation, but I will give it for the sake of completeness. The form of the solution depends on the sign of Ω_c. (It is actually an analytic function of Ω_c with no singularity at $\Omega_c = 0$; however, the analytic continuation is not obvious at a glance.) For an open universe ($\Omega_c > 0$),

$$\Omega_c H_0 t = Q - \sqrt{\Omega_r} - \frac{\Omega_m}{2\sqrt{\Omega_c}}\ln\left\{\frac{\Omega_m + 2\sqrt{\Omega_c}Q + 2\Omega_c y}{\Omega_m + 2\sqrt{\Omega_c\Omega_r}}\right\}, \tag{A.7}$$

where $y = R/R_0$ and $Q = \sqrt{\Omega_r + \Omega_m y + \Omega_c y^2}$. For a flat universe ($\Omega_c = 0$),

$$\Omega_m^2 H_0 t = 2\Omega_m Qy - \tfrac{4}{3}[Q^3 - \Omega_r^{3/2}]. \tag{A.8}$$

A closed universe ($\Omega_c < 0$) expands and then contracts, giving two values of t for each value of y. Following Weinberg's treatment [40] of the $\Omega_r = 0$ case, I will define a development angle $\theta(y)$ by

$$\sin\theta = \frac{2\sqrt{-\Omega_c}Q}{\sqrt{\Omega_m^2 - 4\Omega_r\Omega_c}}. \tag{A.9}$$

As y grows from zero to its maximum value (where $Q = 0$) and then decreases, θ grows from $\theta_0 \equiv \theta(y = 0)$ to $2\pi - \theta_0$. Then

$$2(-\Omega_c)^{3/2} H_0 t = \Omega_m(\theta - \theta_0) - \sqrt{\Omega_m^2 - 4\Omega_r\Omega_c}\,(\sin\theta - \sin\theta_0). \tag{A.10}$$

To understand the horizon problem, we must map out light trajectories in this metric. For this purpose it is useful to rewrite the Robertson-Walker metric (2.1) using the radial variable ξ defined by

$$r(\xi) = \begin{cases} \sin\xi & \text{if} \quad k = 1 \\ \xi & \text{if} \quad k = 0 \\ \sinh\xi & \text{if} \quad k = -1. \end{cases} \tag{A.11}$$

The metric becomes

$$ds^2 = -dt^2 + R^2(t)[d\xi^2 + r^2(\xi)(d\theta^2 + \sin^2\theta d\phi^2)]. \tag{A.12}$$

The coordinate distance which light travels (along a radial line) between times t_1 and t_2 is then given by

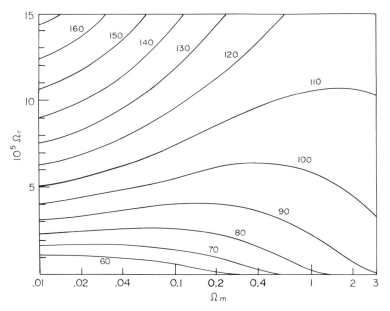

Figure 1

The cosmic microwave background radiation arriving from two opposite directions in the sky comes from sources which were separated from each other at the time of emission by N horizon lengths. This figure shows contour lines of N as a function of Ω_m and Ω_r. Ω_m denotes the ratio of the current mass density due to massive particles to the critical mass density, and Ω_r describes the corresponding ratio for massless particles. Note that $\Omega_r = 4 \times 10^{-5} \tau_0^{\,4} h_0^{-2}$.

$$\xi(t_2, t_1) = \int_{t_1}^{t_2} \frac{dt}{R(t)}. \tag{A.13}$$

The argument in section II can be repeated, with the result that the sources of the microwave background radiation from two opposite directions were separated from each other at the time of emission by N horizon lengths, where

$$N = 2\left[\frac{\xi_H(t_0) - \xi_H(t_r)}{\xi_H(t_r)}\right] \tag{A.14}$$

and $\xi_H(t) \equiv \xi(t, 0)$. Using (A.6), one has

$$\xi_H(t) = \frac{1}{R_0 H_0}\int_0^{y(t)} \frac{dy'}{\sqrt{\Omega_r + \Omega_m y' + \Omega_c y'^2}}, \tag{A.15}$$

where again $y = R/R_0$. Let

$$P(y) \equiv \sqrt{\Omega_r + \Omega_m y + \Omega_c y^2} + \sqrt{\Omega_r}. \tag{A.16}$$

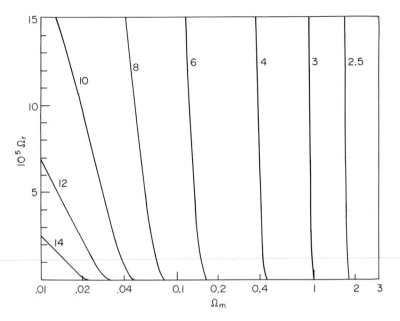

Figure 2
This figure shows contour lines for z_{max}, the maximum value of the red-shift parameter z,
which describes sources that were within one horizon length from us at the time of
emission.

Then

$$R_0 H_0 \xi_H = \begin{cases} \dfrac{1}{\sqrt{\Omega_c}} \ln\left[\dfrac{P + y\sqrt{\Omega_c}}{P - y\sqrt{\Omega_c}}\right] & \text{if } \Omega_c > 0 \\[2ex] \dfrac{2y}{P} & \text{if } \Omega_c = 0 \\[2ex] \dfrac{2}{\sqrt{-\Omega_c}} \tan^{-1}\left[\dfrac{y\sqrt{-\Omega_c}}{P}\right] & \text{if } \Omega_c < 0. \end{cases} \qquad (A.17)$$

Values of N as a function of Ω_r and Ω_m are shown in figure 1. The
graph assumes that recombination occurs at $y = 2.7°\text{K}/4000°\text{K}$. As can
be seen, the value of $N \approx 75$ that was found in section II (corresponding
to $\Omega_r = 0$, $\Omega_m = 1$) is slightly lower than the value of N for more believable
values of Ω_r and Ω_m. One can be quite confident that $\Omega_r > 4 \times 10^{-5}$, in
which case $N > 90$.

An alternative way to illustrate the horizon problem is to ask the
following question: how far can we look out into space such that the
object we are looking at was within one horizon distance from us at the

time of emitting the light? The answer to this question is shown in figure 2. Here distances are expressed in terms of the red-shift parameter z, defined by

$$z = \frac{R_0}{R} - 1 = \frac{1}{y} - 1.$$ (A.18)

Recombination at a temperature of $4000°K$ corresponds to $z \approx 1500$. Figure 2, on the other hand, shows that the largest possible value of z which is consistent with causal contact would be $z \approx 13$, for the case $\Omega_r \approx 4 \times 10^{-5}$ and $\Omega_m \approx 0.01$.

Acknowledgments

I first want to thank Henry Tye (who was also a student of Francis's), who is solely responsible for having persuaded me that the early universe is worthy of study. I also want to thank Sidney Coleman and Lenny Susskind, who provided much encouragement and advice when the inflationary scenario was taking shape. I must also thank Erick Weinberg, who helped me explore the consequences of the scenario. Finally, I would like acknowledge very useful conversations with many people, including Michael Aizenman, Harry Kesten, Paul Langacker, Gordon Lasher, So-Young Pi, John Preskill, Gary Steigman, Paul Steinhardt, and Edward Witten. This work was supported in part by the U.S. Department of Energy under contract DE-ACO2-76ERO3069 and by an Alfred P. Sloan Foundation Followship.

References

[1] M. Gell-Mann, M. L. Goldberger, N. M. Kroll, and F. E. Low, *Phys. Rev.* 179: 1518 (1969).

[2] S. Weinberg, *Phys. Rev. Lett.* 19: 1264 (1967).

[3] S. L. Glashow, J. Iliopoulos, and L. Maiani, *Phys. Rev. D* 2: 1285 (1970).

[4] For an excellent review of grand unified theories, see P. Langacker, *Phys. Rep. C* 72: 185 (1981).

[5] H. Georgi, H. R. Quinn, and S. Weinberg, *Phys. Rev. Lett.* 33: 451 (1974).

[6] T. J. Goldman and D. A. Ross, *Phys. Lett. B* 84: 208 (1979) and *Nucl. Phys. B* 171: 273 (1980).

[7] H. Georgi and S. L. Glashow, *Phys. Rev. Lett.* 32: 438 (1974).

[8] See S. Gasiorowicz and J. L. Rosner, *Am. J. Phys.* 49: 954 (1981) and references therein.

[9] A. D. Sakharov, *Zh ETF Pis. Red.* 5: 32 (1967) [*JETP Lett.* 5: 24 (1967)]; M. Yoshimura, *Phys. Rev. Lett.* 41: 281 (1978); S. Dimopoulos and L. Susskind, *Phys. Rev. D* 18:

4500 (1978); D. Toussaint, S. B. Treiman, F. Wilczek, and A. Zee, *Phys. Rev. D* 19: 1036 (1979); and S. Weinberg, *Phys. Rev. Lett.* 42: 850 (1979). For a more complete list of references, see refs. [4] and [14].

[10] S. Barr, G. Segre, and H. A. Weldon, *Phys. Rev. D* 20: 2494 (1979).

[11] D. V. Nanopoulos and S. Weinberg, *Phys. Rev. D* 20: 2484 (1979).

[12] E. P. Tryon, *Nature* 246: 396 (1973); R. Brout, F. Englert, and P. Spindel, *Phys. Rev. Lett.* 43: 417 (1979); D. Atkatz and H. Pagels, *Phys. Rev. D* 25: 2065 (1982); and J. R. Gott, *Nature* 295: 304 (1982).

[13] See E. Witten, *Comm. Math. Phys.* 80: 381 (1981) and references therein.

[14] A. H. Guth, *Phys. Rev. D* 23: 347 (1981).

[15] M. Aaronson, J. Mould, J. Huchra, W. T. Sullivan, R. A. Schommer, and G. D. Bothun, *Ap. J.* 239: 12, 66-B1 (1980); D. Branch, *Mon. Not. R. Astron. Soc.* 186: 609 (1979); G. de Vacouleurs and G. Bollinger, *Ap. J.* 233: 433 (1979); and A. Sandage and G. A. Tammann, *Ap. J.* 210: 7 (1976).

[16] D. N. Schramm and G. Steigman, *Ap. J.* 243: 1 (1981); S. M. Faber and J. S. Gallagher, *Ann. Rev. Astr. Ap.* 17: 135 (1979); and P. J. E. Peebles, *Astron. J.* 84: 730 (1979).

[17] J. Yang, D. N. Schramm, G. Steigman, and R. T. Rood *Ap. J.* 227: 697 (1979).

[18] D. A. Kirzhnits and A. D. Linde, *Phys. Lett. B* 42: 471 (1972); S. Weinberg, *Phys. Rev. D* 9: 3357 (1974); L. Dolan and R. Jackiw, *Phys. Rev. D* 9: 3320 (1974); A. D. Linde, *Rep. Prog. Phys.* 42: 389 (1979); A. H. Guth and S.-H. Tye, *Phys. Rev. Lett.* 44: 631, 963 (1980); M. B. Einhorn, D. L. Stein, and D. Toussaint, *Phys. Rev. D* 21: 3295 (1980); M. Daniel and C. E. Vayonakis, *Nucl. Phys. B* 180: 301 (1981); A. H. Guth and E. J. Weinberg, *Phys. Rev. D* 23: 876 (1981).

[19] J. P. Preskill, *Phys. Rev. Lett.* 43: 1365 (1979). See also Y. B. Zeldovich and M. Y. Khlopov, *Phys. Lett. B* 79: 239 (1978).

[20] P. Langacker and S.-Y. Pi, *Phys. Rev. Lett.* 45: 1 (1980).

[21] D. N. Schramm and E. M. D. Symbalisty, *Rep. Prog. Phys.* 44: 293 (1981).

[22] P. J. E. Peebles, *Ap. J.* 153: 1 (1968).

[23] See, e.g., A. H. Guth and S.-H. Tye, ref. [18].

[24] S. W. Hawking and G. F. R. Ellis, *The Large Scale Structure of Space-Time*, Cambridge Univ. Press, Cambridge (1973).

[25] J. D. Barrow and M. S. Turner, *Nature* 292: 35 (1981).

[26] G. Steigman, to appear in the proceedings of the Europhysics Study Conference: Unification of the Fundamental Interactions—II, Erice, 6–14 October 1981.

[27] J. Ellis and G. Steigman, *Phys. Lett. B* 89: 186 (1980).

[28] A. H. Guth and E. J. Weinberg, ref. [18].

[29] S. Coleman, *Phys. Rev. D* 15: 2929 (1977); C. G. Callan and S. Coleman, *ibid.* 16: 1762 (1977); S. Coleman, "The Whys of Subnuclear Physics," in *Proceedings of the International School of Subnuclear Physics, Ettore Majorana, Erice, 1977*, edited by A. Zichichi, Plenum, New York (1979).

[30] A. H. Guth and E. J. Weinberg, "Could the Universe Have Recovered from a Slow First-Order Phase Transition?," MIT preprint CTP #950 (Feb., 1982), submitted to *Nucl. Phys. B*.

[31] S. Coleman and E. J. Weinberg, *Phys. Rev. D* 7: 1888 (1973).

[32] L. F. Abbott, *Nucl. Phys. B* 185: 233 (1981); A. Billoire and K. Tamvakis, *Nucl. Phys.*

B 200 [FS4]: 329 (1982); M. Sher, *Phys. Rev. D* 24, 1699 (1981); and G. P. Cook and K. T. Mahanthappa, *Phys. Rev. D* 25: 1154 (1982).

[33] A. Albrecht and P. J. Steinhardt, *Phys. Rev. Lett.* 48: 1220 (1982)

[34] A. D. Linde, *Phys. Lett. B* 108: 389 (1982).

[35] G. Gibbons and S. W. Hawking, *Phys. Rev. D* 17: 2738 (1977).

[36] A. Vilenkin, "Gravitational Effects in Guth Cosmology," Tufts Univ. preprint (1982).

[37] W. H. Press, in *Cosmology and Particles, Proceedings of the Recontres de Moriond, 1981,* edited by J. Audouze, P. Crane, T. Gaisser, D. Hegyi, and J. Tran Thanh Van, Editions Frontieres, Gif-sur-Yvette, France.

[38] S. Coleman and F. De Luccia, *Phys. Rev. D* 21: 3305 (1980).

[39] E. Farhi, private communication.

[40] S. Weinberg, *Gravitation and Cosmology*, Wiley, New York (1972), p. 482.

[41] S. W. Hawking and I. Moss, *Phys. Lett. B* 110: 35 (1982).

Cosmological Consequences of Grand Unified Theories

So-Young Pi

Early-universe phase transitions that are implied by grand unified theories can suppress the primordial monopole density. These are described and the possibility of baryon number generation in this context is discussed.

I. Introduction

In the last few years, there has been a growing relationship between particle physics and cosmology due to the development of grand unified theories [1] of strong, weak, and electromagnetic interactions. Other than proton decay, new physics in these models occurs at very high energies ($\gtrsim 10^{14}$ Gev), which were attained only in the early universe. Consequently both particle physicists and cosmologists are using the same laboratory to test their ideas. The former are explaining some aspects of cosmology and the latter in return are providing constraints on elementary-particle physics.

Grand unified theories are based on the assumption that there exists a simple group G which is characterized by a single coupling constant g, and that the symmetries of strong, weak, and electromagnetic interactions are contained in G. One supposes that at some very high mass scale, $M_X \gtrsim 10^{14}$ Gev, G breaks down to $SU(3)^c \times SU(2) \times U(1)$, which is the group of the above interactions. At an energy scale $M_W \approx 100$ GeV the weak and electromagnetic gauge group, $SU(2) \times U(1)$, further breaks down to $U(1)^{em}$:

$$G \to SU(3)^c \times SU(2) \times U(1) \to SU(3)^c \times U(1)^{em}. \qquad (1.1)$$
$$\uparrow \qquad\qquad\qquad\qquad \uparrow$$
$$M_X \gtrsim 10^{14}\ \text{Gev} \qquad M_W \approx 100\ \text{GeV}$$

Theories based on the above symmetry behavior in general have two consequences, which play an important role in the interconnection between grand unified theories and cosmology. Firstly, there exist baryon-number-nonconserving interactions. Quarks and leptons form a single multiplet of G and therefore G's generators necessarily connect them. This leads to violation of baryon and lepton number conservation. The baryon-number-nonconserving forces, together with C- and CP-violating interactions, may provide a theory in which the baryon asymmetry of the universe appears dynamically [2]. Secondly, the symmetry behavior of (1.1) implies the existence of superheavy magnetic monopoles. These monopoles are of the 't Hooft-Polyakov type [3] that occur when any simple group is spontaneously broken to a subgroup containing a $U(1)$ factor. Their mass M_m is of order M_X/α, where M_X is a typical mass of the gauge bosons associated with a broken generator, and $\alpha = g^2/4\pi$ is the fine-structure constant of the group. In grand unified theories M_m is typically $O(10^{16}\text{ GeV})$. If these monopoles were produced prolifically shortly after the big bang they would contribute too much to the energy density of the universe [4]. In the following I shall summarize the baryon asymmetry of the universe and initial production of superheavy magnetic monopoles.

Baryon Asymmetry of the Universe
A fundamental fact about our universe is that it is predominantly matter, not an equal mixture of matter and antimatter, at least in the local region of our galaxy [5]. The matter content of the present universe is described by the ratio of baryon number density $n_B (= $ baryons $-$ antibaryons), to photon number density n_γ [6]:

$$n_B/n_\gamma \approx 10^{-10\pm1}. \tag{1.2}$$

In order to have a theory which explains this fundamental parameter of cosmology dynamically, the following requirements must be satisfied:

(a) baryon-nonconserving forces must be present,

(b) C- and CP-violating interactions must occur,

(c) the universe must have been out of thermal equilibrium.

Baryon-nonconserving forces exist in grand unified theories and the requirement (b) can be easily incorporated. The third requirement is necessary because in thermal equilibrium no asymmetry between particles and antiparticles develops. Departures from thermal equilibrium may have occurred in an expanding universe. But there are subtleties: light particles tend to keep their equilibrium distribution even in a situation which is far from equilibrium. Therefore it is difficult to generate a net baryon number from interactions involving only light particles [7, 8]. This suggests that we need heavy particles. The most promising candidates that can generate baryons are the decays of heavy vector bosons and scalars.

Let us consider a heavy boson with mass M_B. Its decay will occur with a rate $\Gamma \approx \alpha M_B$. The production of the boson by inverse decay at temperature T occurs with a rate αM_B ($M_B \lesssim T$) or $\alpha M_B \exp\{-M_B/T\}$ ($M_B \gtrsim T$). The characteristic time scale is set by the expansion rate of the universe $\gamma \approx T^2/M_P$ where M_P is the Planck mass, 10^{19} GeV. As T decreases, if the decay rate is smaller than the expansion rate, the distribution will be out of equilibrium. Alternatively if the decay proceeds faster than the expansion rate, then the heavy-boson number density will decrease to an equilibrium distribution. By comparing the decay rate and the expansion rate, we obtain a temperature T_D above which $\Gamma < \gamma$ and below which $\Gamma > \gamma$:

$$\alpha M_B \approx T_D^2/M_P. \tag{1.3}$$

The net frozen number of bosons, which existed at $T \gtrsim T_D$ due to a non-equilibrium distribution, will decay faster than the expansion rate as the temperature decreases below T_D. In this way an equilibrium distribution will be achieved and net baryon number can be created. In order to forbid the inverse decay, we require $T_D \lesssim M_B$. This, together with (1.3), gives a bound for the mass of heavy particles and the temperature at which baryon number generation occurs:

$$M_B \gtrsim \alpha M_P, \quad T_D \lesssim M_B. \tag{1.4}$$

In grand unified theories, with plausible and general assumptions, the above decay mechanism yields a value of n_B/n_γ in the early universe that is close to the present value. A simple explanation for the currently observed baryon to photon ratio can now be given, if the expansion of the universe has been adiabatic: n_B/n_γ is a relic.

However, this scenario does not work if the universe underwent any highly nonequilibrium processes between the time of baryon synthesis and the present, such as strongly first-order phase transitions with supercooling [9] or some other entropy-generating reactions [10]. In that case n_B/n_γ would change by a large factor.

Initial Monopole Production
The problem of monopole production and their subsequent annihilation,

in the context of a second-order phase transition, was analyzed by Zeldovich and Khlopov as well as by Preskill [4].

It is well known that in spontaneously broken gauge theories the symmetry is fully restored, in general, at very high temperatures [11]. In the standard hot big-bang cosmology, the temperature of the universe exceeded the critical temperature; therefore initially the symmetry was unbroken. As the universe expands and cools, it undergoes a phase transition and the symmetry will be broken. In grand unified theories we expect

$$G \to SU(3)^c \times SU(2) \times U(1). \tag{1.6}$$
$$\uparrow$$
$$T_c \gtrsim M_X$$

Near the critical temperature T_c monopoles are produced copiously due to large fluctuations in the Higgs fields. They then annihilate strongly, when they are close together. However, as the universe expands, the annihilations become negligible. Preskill's estimate for the monopole density to entropy ratio, at temperatures well below the critical temperature, is

$$\frac{n_m}{n_\gamma} \approx 10^{-6} \frac{M_m}{M_P}. \tag{1.7}$$

This tells us that the annihilation can reduce n_m/n_γ only to $O(10^{-10})$ for $M_m \approx 10^{16}$ GeV. However, the standard scenario of helium synthesis requires $n_m/n_\gamma \lesssim O(10^{-19})$ at $T = 1$ MeV. Moreover, the bound on the present value of n_m/n_γ imposed by the observed Hubble constant and deceleration parameter is $\lesssim 10^{-24}$. An important cosmological constraint on grand unified theories therefore is that the initial monopole density must be strongly suppressed, so that the present density does not exceed the established bounds.

Possible solutions to this problem [12–14] depend, in general, on the phase transitions in the early universe. For each such transition baryon synthesis should be reexamined, since it also depends on the thermal history of the universe. In the following I shall describe phase transitions which can suppress initial monopole production and discuss their cosmological consequences, including baryon production.

II. Suppression of Initial Monopoles in Second-Order Phase Transitions

We shall describe a simple solution [14] to the problem of initial monopole overabundance, in which the universe undergoes two or more phase transitions that can be second order:

$$G \to H_1 \to H_2 \to \cdots \to H_n \to SU(3)^c \times U(1)^{em}. \qquad (2.1)$$
$$\quad T_1 \quad T_2 \qquad\qquad T_c$$

Here, the critical temperature T_c at which $U(1)^{em}$ appears is $\gtrsim 1$ TeV. $U(1)^{em}$ is not a factor of H_n. $U(1)^{em}$ is either spontaneously broken for $T > T_c$ or it is contained in H_n. While the latter possibility would not produce heavy monopoles, it is unacceptible because it would, in general, imply an extremely short lifetime for the proton, which could decay via the interactions in H_n. We choose the first possibility and propose that $U(1)^{em}$ is spontaneously broken for $T \gtrsim T_c$. We are therefore departing from the generic situation of symmetry increase above a critical temperature. Rather in our model, above the critical temperature T_c the symmetry is decreased. While this is unusual, it can happen in special circumstances [15]. For example, G could break down to $SU(3)^c$ at $T \gtrsim M_X$ and undergo a second-order phase transition to $SU(3)^c \times U(1)^{em}$ at $T_c \gtrsim 1$ TeV. In this case, since $T_c \ll M_m \approx 10^{16}$ GeV, no monopoles will be produced.

An explicit model which exhibits this behavior will now be described. At $T = 0$ it is the standard $SU(5)$ theory with $SU(5)$ broken to $SU(3)^c \times SU(2) \times U(1)$ by an adjoint Higgs representation and then to $SU(3)^c \times U(1)^{em}$ by one or more fundamental representations of the Higgs fields:

$$SU(5) \to SU(3)^c \times SU(2) \times U(1) \to SU(3)^c \times U(1)^{em}. \qquad (2.2)$$
$$\quad\uparrow \qquad\qquad\qquad\qquad\quad \uparrow$$
$$M_X \gtrsim 10^{14} \text{ GeV} \qquad\quad M_W \approx 10^2 \text{ GeV}$$

But at high temperature we want the following symmetry behavior:

$$SU(5) \quad\to\quad SU(3)^c \quad\to\quad SU(3)^c \times U(1)^{em}. \qquad (2.3)$$
$$\quad\uparrow \qquad\qquad \uparrow$$
$$T \gtrsim M_X \qquad T_c \gtrsim 1 \text{ TeV}$$

The $SU(2) \times U(1)$ electroweak group is completely broken for $T \gtrsim T_c$, but $U(1)^{em}$ is restored for $T \lesssim T_c$.

Let us briefly demonstrate the possibility of such a symmetry behavior. For $0 \le T \ll M_X$, we only need to consider the $SU(2) \times U(1)$ part of the model. (The heavy-particle contributions to the finite temperature potential are negligible [11].) We shall consider a $SU(2) \times U(1)$ model with three Higgs doublets. This is the minimum number of Higgs fields required for our purpose. The Higgs potential $V(\phi)$ at $T = 0$ is

$$V(\phi) = \sum_{i=1}^{3} \left[-\mu_i^2 \phi_i^\dagger \phi_i + \lambda_i (\phi_i^\dagger \phi_i)^2 \right]$$
$$+ \sum_{i<j} \left[\sigma_{ij} \phi_i^\dagger \phi_i \phi_j^\dagger \phi_j + \rho_{ij} \phi_i^\dagger \phi_j \phi_j^\dagger \phi_i \right. \tag{2.4}$$
$$\left. + \eta_{ij} (\phi_i^\dagger \phi_j)^2 + \eta_{ij}^* (\phi_j^\dagger \phi_i)^2 \right],$$

where we have imposed discrete symmetries $\phi_i \to -\phi_i$ for simplicity. If the minimum of V occurs when only one doublet has a nonzero vacuum expectation value (VEV), e.g., $\langle \phi_1 \rangle \neq 0$, then

$$\langle \phi_1 \rangle = \frac{1}{\sqrt{2}} \begin{bmatrix} 0 \\ v_1 \end{bmatrix} \tag{2.5}$$

and $SU(2) \times U(1)$ is broken to $U(1)^{\text{em}}$. If two doublets ϕ_1 and ϕ_2 have nonzero VEVs, then either

$$\langle \phi_1 \rangle = \frac{1}{\sqrt{2}} \begin{bmatrix} 0 \\ v_1 \end{bmatrix}, \quad \langle \phi_2 \rangle = \frac{1}{\sqrt{2}} \begin{bmatrix} v_2 \\ 0 \end{bmatrix}, \quad \langle \phi_3 \rangle = 0, \tag{2.6}$$

or

$$\langle \phi_1 \rangle = \frac{1}{\sqrt{2}} \begin{bmatrix} 0 \\ v_1 \end{bmatrix}, \quad \langle \phi_2 \rangle = \frac{1}{\sqrt{2}} \begin{bmatrix} 0 \\ v_2 \end{bmatrix}, \quad \langle \phi_3 \rangle = 0, \tag{2.7}$$

depending whether ρ_{12} is greater or less than $2|\eta_{12}|$, respectively. In this case $SU(2) \times U(1)$ is either completely broken [eq. (2.6)], or broken to $U(1)^{\text{em}}$ [eq. (2.7)]. We shall take

$$\rho_{ij} > 2|\eta_{ij}|, \tag{2.8}$$

so that if several fields get nonzero VEVs they tend to be orthogonal at the minimum, as in (2.6).

First, we choose the parameters such that $\langle \phi_1 \rangle_{T=0}$ has the form (2.5) and $\langle \phi_2 \rangle_{T=0} = \langle \phi_3 \rangle_{T=0} = 0$. Then at $T = 0$, $SU(2) \times U(1)$ is broken to $U(1)^{\text{em}}$. This occurs for

$$\mu_1^2 > 0, \quad \mu_{2,3}^2 < 0,$$
$$|\mu_i|^2 + (\sigma_{1i} \mu_1^2 / 2\lambda_1) > 0, \quad \text{for} \quad i = 2, 3. \tag{2.9}$$

We also require

$$\lambda_i > 0, \quad \sigma_{ij} > -\sqrt{\lambda_i \lambda_j}, \tag{2.10}$$

which with (2.8) is sufficient to guarantee that V is bounded from below.

For $T > 0$, we have to find the minimum $\langle \phi_i \rangle_T$ by minimizing the finite-temperature effective potential $V(\phi, T)$. For sufficiently high temperatures, $V(\phi, T)$ is given by [11]:

$$V(\phi, T) = V(\phi, T = 0) - \frac{\pi^2}{90} N(T) T^4 + \sum_{i=1}^{3} \frac{1}{2} T^2 F_i \phi_i^\dagger \phi_i, \qquad (2.11)$$

where $N(T) = N_b(T) + \frac{7}{8} N_f(T)$, and N_b and N_f are the total number of helicity states for bosons and fermions of mass $\ll T$. The quantities F_i are

$$F_i = \frac{3g^2 + g'^2}{8} + \lambda_i + \sum_{i \neq j} \frac{\sigma_{ij}}{3} + \frac{\rho_{ij}}{6} + \text{Yukawa terms.} \qquad (2.12)$$

For small fermion masses, the Yukawa terms in (2.12) will generally be negligible. $V(\phi, T)$ has effective mass terms

$$M_i^2(T) = -\mu_i^2 + \tfrac{1}{2} F_i T^2. \qquad (2.13)$$

If F_i are positive, then for $T^2 \gtrsim 2\mu_1^2/F_1$, the effective mass $M_1^2(T)$ becomes positive ($M_{2,3}^2(T)$ are positive at all T) and the system will undergo a transition to a phase in which $SU(2) \times U(1)$ is unbroken ($\langle \phi_1 \rangle_T = 0$). However, some of the F_i can be negative [15], and the symmetry need not be restored at high temperature; indeed it can be lowered. We choose parameters so that $F_{1,2} < 0$ and $V(\phi, T)$ has a minimum of the form of (2.6) at high temperatures:

$$\lambda \equiv \lambda_1 \approx \lambda_2 \qquad \lambda \gg g^4, |\rho_{ij}|$$

$$-\sigma \equiv \sigma_{13} \approx \sigma_{23} \qquad \sigma > 3\lambda + \sigma_{12} + 3X \qquad (2.14)$$

$$\sigma_{12}/\lambda < 2 \qquad |\mu_2^2| > \mu_1^2$$

where $X = (3g^2 + g'^2)/8$. The condition $\lambda \gg g^4$ allows neglecting radiative corrections to V. For the above parameters, $F_3 \approx \lambda_3 > 0$, $|F_1| \approx |F_2| \lesssim O(\lambda \approx g^2)$, and $V(\phi, T)$ has a minimum of the form of (2.6) with

$$\begin{bmatrix} v_1(T)^2 \\ v_2(T)^2 \end{bmatrix} = (\lambda_1 \lambda_2 - \sigma_{12}^2/4)^{-1} \begin{bmatrix} \lambda_2 & -\sigma_{12}/2 \\ -\sigma_{12}/2 & \lambda_1 \end{bmatrix} \begin{bmatrix} -M_1^2(T) \\ -M_2^2(T) \end{bmatrix}. \qquad (2.15)$$

We find that there is a range of parameters that satisfy conditions (2.8), (2.9), (2.10), and (2.14). The gauge symmetry $SU(2) \times U(1)$ is completely broken at this minimum. The phase transition from the low-temperature $U(1)^{\text{em}}$ phase to this minimum occurs at

$$T_c = A\mu_1/\sqrt{\lambda} \approx A(246 \text{ GeV}). \qquad (2.16)$$

A is a function of parameters in the potential and is typically of order unity. But it can be made larger or smaller by adjusting parameters. We shall assume $T_c \gtrsim 1$ TeV.

We have therefore demonstrated the existence of a model for which $SU(2) \times U(1)$ is completely broken for $T \gtrsim T_c$ but $U(1)^{em}$ is restored for $T \lesssim T_c$.

This $SU(2) \times U(1)$ model can easily be embedded in $SU(5)$ [16]. For $T \gtrsim M_X$ the superheavy particles can no longer be neglected and their contributions must be added to the effective potential. It has been shown that $SU(5)$ symmetry can be restored for $T > M_X$. There may also be intermediate phases (e.g., $SU(3)^c \times SU(2) \times U(1)$) for $T \lesssim M_X$:

$$SU(5) \quad \rightarrow \quad \text{possible intermediate phases}$$
$$T \gtrsim M_X$$

$$\rightarrow \quad SU(3)^c \rightarrow SU(3)^c \times U(1)^{em}. \quad (2.17)$$
$$T < M_X \qquad T_c$$

This unconventional pattern of finite-temperature symmetry behavior has the following cosmological consequences.

(a) Stable monopoles of mass $M_m \approx 10^{16}$ GeV could, in principle, exist for $T < T_c$. But their number density, expected from thermal fluctuation to be $\approx \exp(-M_m/T_c)$, is utterly negligible. Once the $SU(3)^c$ phase is entered, monopoles produced during intermediate phases will be confined in pairs that could subsequently annihilate.

(b) An interesting feature is that electric-charge conservation is violated for $T \gtrsim T_c$. Charge-violating reactions occur through Yukawa interactions, and they are found to occur with a typical rate

$$\Gamma(T) \approx f^2 T. \quad (2.18)$$

$\Gamma(T)$ is large compared to γ, the expansion rate of the universe ($\Gamma/\gamma \gtrsim 10^7$ GeV/T), and therefore charge-violating reactions are in equilibrium.

(c) Particle masses which are light ($\lesssim 100$ GeV) at $T = 0$ are proportional to the temperature for $T \gtrsim T_c$. The gauge boson ($W_{\pm}, Z, \gamma$) masses are $M_V \approx gT$, but are negligible compared to the electron plasma frequency

$$\omega_p(T) \approx 400\ T \quad (2.19)$$

and can be ignored. Fermion masses are

$$m_F(T) \approx m_F(0)\sqrt{G_F}T \ll T. \quad (2.20)$$

They are sufficiently small not to affect baryon production. The simple scenario for baryon production described in section I can occur in this model shortly after the phase transition,

$$SU(5) \quad \rightarrow \quad SU(3)^c \times SU(2) \times U(1)$$
$$T \gtrsim M_X$$

or   (2.21)

$$SU(5) \quad \rightarrow \quad SU(3)^c.$$
$$T \gtrsim M_X$$

Thus monopoles are suppressed, without affecting the conventional baryon-production scenario.

III. Suppression of Initial Monopole Density by a Strongly First-Order Phase Transition

An alternative mechanism for suppressing monopoles in the early universe makes use of a first-order phase transition with supercooling [12]. Grand unified theories may undergo transitions

$$G \quad \rightarrow \quad G' \rightarrow SU(3)^c \times SU(2) \times U(1) \qquad (3.1)$$
$$\uparrow$$
$$T_c \approx 10^{14} \text{ GeV} \approx M_X$$

where G' is a metastable phase. The transition from G to G' at T_c may be second order. That from G' to $SU(3)^c \times SU(2) \times U(1)$ is first order. This is a process in which a finite region of space tunnels through a barrier in the potential to form a bubble of the new phase [17]. Then the bubbles grow and coalesce to form a larger space. Kibble [18] has pointed out that monopoles are produced by the uncorrelated Higgs fields inside bubbles when they coalesce. Consequently, the number of monopoles produced in this way is comparable, within a few orders of magnitude, to the number of bubbles. Therefore, small-bubble nucleation rates can suppress the initial monopole density.

Using this picture, Guth and Tye [12] estimated a bound on monopole density at the temperature of bubble coalescence:

$$n_M > 10^4 T_{\text{coal}}^6 / M_P^3. \qquad (3.2)$$

In order to satisfy the bound $n_m/n_\gamma \lesssim 10^{-24}$, T_{coal} must be less than $O(10^{11} \text{ GeV})$. This implies that if $T_c \approx 10^{14}$ GeV, the universe must supercool by at least three orders of magnitude in order to suppress the initial monopole density.

An explicit calculation [12] in the $SU(5)$ model with Higgs fields in the adjoint representation, shows that there is a phase transition

$$SU(5) \quad \rightarrow \qquad SU(4) \times U(1)' \rightarrow SU(3)^c \times SU(2) \times U(1) \qquad (3.4)$$
$$\uparrow$$
$$T_c \approx 10^{14} \text{ GeV}$$

where the $SU(4) \times U(1)'$ phase is metastable. The amount of supercooling in the $SU(4) \times U(1)'$ phase depends on the nucleation rate of bubbles, which again depends on the undetermined parameters in the potential. Sufficient supercooling can be obtained by adjusting parameters [19]. However, one cannot ignore the effects of monopole-like objects in the $SU(4) \times U(1)'$ phase which could be produced in large numbers at T_c. In fact, these monopole-like objects in the metastable vacuum can affect the first-order phase transition significantly [20]. But if we study the high-temperature behavior of the complete Higgs potential, including the fundamental representation Higgs fields, we find that for a wide range of parameters [21]:

$$SU(5) \rightarrow SU(4) \times U(1)' \rightarrow SU(4) \rightarrow SU(3)^c \times SU(2) \times U(1). \qquad (3.5)$$

Therefore the monopoles in $SU(4) \times U(1)'$ disappear as the $SU(4)$ phase enters, and supercooling may occur in a $SU(4)$ metastable phase.

The first-order phase transition with supercooling not only solves the monopole problem but has other interesting consequences. When the universe supercools until $T \ll T_c$, its expansion becomes exponential in time and no longer adiabatic. Guth [22] has argued that if the universe supercooled to temperatures 28 or more orders of magnitude below the critical temperature, the two initial difficulties with the standard model of hot big-bang cosmology—the horizon and flatness problems—would disappear.

But this scenario is incomplete [22, 23]. In order to have sufficient supercooling to solve the monopole, horizon, and flatness problems, the bubble nucleation rate must be smaller than the fourth power of the expansion rate of the universe. One of the effects of this random bubble formation is that the latent heat released after the phase transition cannot be thermalized. The heat released as the bubbles expand is transferred initially to their walls and this can be thermalized only when the bubble walls undergo many collisions. If bubble formation is random, the collisions necessary for thermalization cannot occur. As we pointed out already, baryon production must occur after the first-order phase transition, in order that an acceptable n_B/n_γ be maintained. Moreover, the temperature must be near T_c. Therefore the scenario with supercooling is realistic only if a reheating of the universe is accomplished [24].

IV. Suppression of Monopoles in a Hierarchical Supersymmetric Model

Finally, in this last section we shall describe the early universe in a super-symmetric grand unified model in which there is strong suppression of initial monopole production [25]. To solve the mass-hierarchy problem, Witten [26] has proposed a $SU(5)$ supersymmetric model in which a small-mass scale (weak-interaction scale) is fundamental, the large one (unification scale) being generated dynamically.

In the model, there are two complex Higgs fields A_j^i and Y_j^i in the adjoint representation of $SU(5)$, and one complex singlet Higgs field X. The Higgs field in the fundamental representation is ignored for simplicity. The scalar potential is given by

$$
\begin{aligned}
V_0(A, Y, X) = {} & g^2|\operatorname{Tr} A^2 - M^2|^2 + \lambda^2(\operatorname{Tr} A^2 A^{*2} - \tfrac{1}{5}\operatorname{Tr} A^2 \operatorname{Tr} A^{*2}) \\
& + \operatorname{Tr}\{[\lambda(AY + YA) - \tfrac{2}{5}\lambda\operatorname{Tr} AY + 2gXA] \cdot \\
& [\lambda(AY + YA) - \tfrac{2}{5}\lambda\operatorname{Tr} AY + 2gXA]^*\} \\
& + e^2\operatorname{Tr}(i[A, A^*] + i[Y, Y^*])^2,
\end{aligned}
\tag{4.1}
$$

where * denotes complex conjugation.

M is the only mass scale in the theory, and it characterizes super-symmetry breaking. We shall take M to be the weak-interaction scale.

To minimize the energy, one must have

$$
A = \frac{gM}{\sqrt{30g^2 + \lambda^2}}
\begin{bmatrix}
2 & & & & \\
& 2 & & & \\
& & 2 & & \\
& & & -3 & \\
& & & & -3
\end{bmatrix}.
\tag{4.2}
$$

Y is parallel to A at the minimum, and is given by

$$
Y = \frac{g}{\lambda} X
\begin{bmatrix}
2 & & & & \\
& 2 & & & \\
& & 2 & & \\
& & & -3 & \\
& & & & -3
\end{bmatrix}.
\tag{4.3}
$$

However, X is undetermined at the tree level. The broken gauge symmetry is $SU(3) \times SU(2) \times U(1)$. The one-loop effective potential must be calculated in order to determine X; it is given by [27]

$$V_1^0(\phi) = \sum_i \frac{(-1)^F}{64\pi^2} M_i^4(\phi) \ln \{M_i^2(\phi)/\mu^2\}, \tag{4.4}$$

where the sum runs over all helicity states, $M_i(\phi)$ is the field dependent mass of the ith such state, $F = 1$ for fermions, $F = 0$ for bosons, and μ is a renormalization mass. Witten has shown [26] that for large X, $X \gg M$, the effective potential evaluated at the minimum, given by eqs. (4.2) and (4.3), has the following form, which includes the lowest-order term:

$$V_1^0(X) = \frac{M^2 g^2 \lambda^2}{30g^2 + \lambda^2} \left[1 + \frac{g^2}{g^2 + \frac{1}{30}\lambda^2} \frac{(29\lambda^2 - 50e^2)}{80\pi^2} \ln \{|X|^2/\mu^2\} \right]$$
$$+ O\left(\frac{1}{|X|^2}\right) \quad \text{for} \quad X \gg M. \tag{4.5}$$

If $29\lambda^2 - 50e^2 < 0$, then X increases without limit until perturbation theory breaks down. However, Witten argues that asymptotic freedom will force e^2, which really depends on X, to vanish with increasing X. Hence the effective coupling $29\lambda^2 - 50e^2$ changes sign at some large value X_0. Then a stable minimum will be produced near X_0, which can be interpreted as the unification-mass scale ($\gtrsim 10^{14}$ GeV). Therefore, X_0 is independent of the fundamental scale M of the theory and is determined by the renormalization group equation. $SU(5)$ is strongly broken by the vacuum expectation value of Y, according to (4.3).

The interesting early-universe phase transitions which we shall describe below are a consequence of a peculiar feature of this model: the symmetry breaking is entirely determined by the field A which has small-mass characteristics. Therefore, the temperature of the symmetry-changing phase transition is also of the order of the small mass since it is governed by the field A. The finite-temperature potential has the following form. For $T \gg M_i(A, X, Y)$

$$V_{\text{eff}}(T, A, X, Y) = V_0(A, X, Y) - \frac{\pi^2}{90} N(T) T^4 + \sigma_A T^2 \operatorname{Tr} \{AA^*\}$$
$$+ \sigma_Y T^2 \operatorname{Tr} \{YY^*\} + \sigma_X T^2 XX^*, \tag{4.6}$$

where $V_0(A, X, Y)$ is as in (4.1); $N(T)$ is as in (2.11); σ_A, σ_Y, and σ_X are functions of e^2, λ^2, g^2, of order (e^2, λ^2, g^2), and are positive in this model.

For $T \ll M_i(A, X, Y)$

$$V_{\text{eff}}(T, A, X, Y) = V_0(A, X, Y) + V_1^0(A, X, Y), \qquad (4.7)$$

where $V_1^0(A, X, Y)$ is the zero-temperature one-loop potential of (4.4). For regions between $M_i(A, X, Y) \gg T$ and $M_i(A, X, Y) \ll T$, we shall simply extrapolate the two limiting forms (4.6) and (4.7). The above finite-temperature effective potential shows that essentially nothing happens until the temperature becomes the fundamental scale M of the theory. At very high temperature $T \gg X_0$, where X_0 is the large-mass scale ($\gtrsim 10^{14}$ GeV), an absolute minimum exists at $A = X = Y = 0$. The universe is in the $SU(5)$ phase. As the temperature decreases, the field A undergoes a second-order phase transition at

$$T_A = \sqrt{\frac{2}{\sigma_A}} \, gM \approx O(M). \qquad (4.8)$$

For the X, Y fields, the minimum $X = Y = 0$ exists until the temperature becomes $\lesssim O(M)$. In fact, the effective potential of (4.7) shows that there is no minimum at large X and Y as long as the minimum of A is at the origin. Therefore, for $T \gtrsim T_A$, $A = X = Y = 0$ is the only minimum and the universe will stay in the $SU(5)$ phase until the weak-interaction scale M.

For $T \lesssim T_A$, A will have a nonzero vacuum expectation value of the form in (4.2), and the gauge symmetry will be softly broken to $SU(3)^c \times SU(2) \times U(1)$. In addition to the minimum at $X = Y = 0$, another minimum will develop at large X and Y: Y_{min} will have the form of (4.3) and $X_{\text{min}} = X_0$ due to the effective potential (4.5). The minimum at $X = 0$ will become metastable as the temperature decreases and X_0 will become the true vacuum. There are two possible mechanisms through which the phase transition from $X = 0$ to $X = X_0$ can proceed, depending on the sign of $d^2 V_1^0(X)/dX^2|_{X=0}$. The sign depends, in general, on the parameters in the potential. The magnitude of $d^2 V_1^0(X)/dX^2|_{X=0}$ is $\approx M^2 \times (g^4, \lambda^4, e^4)$. We shall describe the two possible cases.

Case I

Let us take $d^2 V_1^0(X)/dX^2|_{X=0} > 0$. Then $d^2 V_{\text{eff}}(X, T)/dX^2|_{X=0}$ is always positive and the transition from $X = 0$ to X_0 will occur through a tunneling process. This is in general a slow process and the temperature T_X at which the tunneling is completed will be well below T_A. As the temperature decreases below T_X, X will increase continuously to X_0.

Case II

Alternatively, if we take $d^2 V_1^0(X)/dX^2|_{X=0} < 0$, then $d^2 V_{\text{eff}}(X, T)/dX^2|_{X=0}$ can be negative and the transition from $X = 0$ to X_0 can proceed without tunneling. If the minimum at $X = 0$ becomes metastable before $d^2 V_{\text{eff}}(X, T)/dX^2|_{X=0}$ becomes negative, tunneling will begin. However, since this is a slow process, the transition will be completed after the curvature becomes negative at some temperature T_X. For $T \lesssim T_X$, X will continuously move from $X = 0$ to X_0.

In both cases, as X increases, the $SU(3)^c \times SU(2) \times U(1)$ symmetry-breaking will become strong.

Let us notice that in this model there is an energy puzzle: There does not seem to be any source for the energy needed to form heavy particles (with mass $\gtrsim 10^{14}$ GeV) because (a) the phase transition at which massive particles first arise occurs at a low temperature, of the order of the weak-interaction scale M, and (b) the energy difference between the vaccua of the two phases $X = 0$, and $X = X_0$ is only $O(M^4)$. One possibility is that no heavy particles will be formed because they decay as X increases to X_0. Of course, this decay cannot occur for magnetic monopoles. Therefore, for them and for any other particles that do not decay, the energy puzzle remains.

Finally let us discuss the cosmological monopole density and baryon production in this model.

(a) Monopole Density When $SU(5)$ breaks down to $SU(3)^c \times SU(2) \times U(1)$ at $T = T_A$ by a second-order phase transition, light monopoles will be produced prolifically. Equation (1.7) tells us that, if T_X is well below T_A, then $n_m/n_\gamma \approx 10^{-6} M_m/M_P$ can be reduced to $O(10^{-20})$. That is, provided monopoles stay light ($M_m \approx O(M)$) long enough, annihilation can reduce n_m/n_γ to the bound required at the time of helium synthesis, even if these monopoles become heavy as X increases to X_0. Therefore, primordial monopole density can be suppressed by adjusting parameters in the potential so that T_X is well below T_A.

(b) Baryon Production If baryons were ever produced, they must have been produced near the phase transition of A and X fields. However, near the critical temperature the expansion rate of the universe will be extremely small and the rate of baryon-generating reactions will be always larger than the expansion rate. Even though there is a possibility that gauge bosons and scalars may decay as X increases, baryon-number generation in this model seems to be an open question.

V. Conclusion and Acknowledgments

The considerations of this essay demonstrate that a profitable relationship exists between particle physics and cosmology: particle physicists can describe the early universe, and constraints from cosmology can discriminate between elementary-particle theories.

I thank my colleagues at Harvard and MIT for discussions, and also the Physics Department of Harvard University for its hospitality. This work was supported by the U.S. Department of Energy under contract ER-78-5-02-4999.

References

[1] The simplest grand unified theory is the $SU(5)$ model of H. Georgi and S. L. Glashow, *Phys. Rev. Lett.* 32: 438 (1974). For other models, see the review by P. Langacker, *Phys. Rep. C* 72: 185 (1981) and references therein.

[2] Yu. Ignatiev, N. V. Krasnikov, V. A. Kuzmin, and A. N. Tavkhelidze, *Phys. Lett. B* 76: 436 (1978); M. Yoshimura, *Phys. Rev. Lett.* 41: 281 (1978); 42: 146 (E) (1979); S. Dimopoulos and L. Susskind, *Phys. Rev. D* 18: 4500 (1978); B. Toussaint, S. B. Treiman, F. Wilczek, and A. Zee, *ibid.* 19: 1036 (1979); J. Ellis, M. K. Gaillard, and D. V. Nanopoulos, *Phys. Lett. B* 80: 360 (1979); S. Weinberg, *Phys. Rev. Lett.* 42: 850 (1979); and A. Yildiz and P. H. Cox, *Phys. Rev. D* 21, 906 (1980). For a review, see M. Yoshimura, in *Grand Unified Theories and Related Topics: Proceedings of the Fourth Kyoto Summer Institute*, edited by M. Konuma and T. Maskawa, World Scientific, Singapore (1981).

[3] G.'t Hooft, *Nucl. Phys. B* 79: 276 (1974); A. M. Polyakov, *Zh. ETF Pis. Red.* 20: 430 (1974) [*JETP Lett.* 20: 194 (1974)].

[4] Y. B. Zeldovich and M. Y. Khlopov, *Phys. Lett. B* 79: 239 (1978); J. P. Preskill, *Phys. Rev. Lett.* 43: 1365 (1979). See also M. B. Einhorn, D. L. Stein, and D. Toussaint, *Phys. Rev. D* 21: 3295 (1980).

[5] See, e.g., T. Weeks, *High Energy Astrophysics*, Chapman and Hall, London (1969).

[6] K. A. Olive, D. N. Schramm, G. Steigman, M. S. Turner, and J. Yang, *Astrophys. J.* 246: 557 (1981).

[7] D. Toussaint, S. B. Treiman, F. Wilczek, and A. Zee, in ref. [2].

[8] S. Weinberg, in ref. [2].

[9] E. Witten, *Nucl. Phys. B* 177: 477 (1981); M. A. Sher, Phys. Rev. *D* 22: 2989 (1980); P. J. Steinhardt, *Nucl. Phys. B* 179: 492 (1981); and A. H. Guth and E. J. Weinberg, *Phys. Rev. Lett.* 45: 1131 (1980).

[10] D. Dicus and V. Teplitz, to be published in *Phys. Rev. Lett.* (These authors discuss suppression of monopole-to-entropy ratio by monopole annihilations in which entropy increases by a large factor.)

[11] L. Dolan and R. Jackiw, *Phys. Rev. D* 9: 3320 (1974); S. Weinberg, *ibid.* 9: 3357 (1974). See also D. A. Kirhznitz and A. D. Linde, *Ann. Phys.* (N.Y.) 101: 195 (1976).

[12] A. H. Guth and S.-H. Tye, *Phys. Rev. Lett.* 44: 631 (1980); 44: 963 (1980); and A. H. Guth and E. Weinberg, *Phys. Rev. D* 23: 876 (1981).

[13] G. Lazarides and Q. Shafi, *Phys. Lett. B* 94: 149 (1980); G. Lazarides, M. Magg, and Q. Shafi, *Phys. Lett. B* 97: 87 (1980) (the suppression discussed here relies on a confinement mechanism); and ref. [10].

[14] P. Langacker and S.-Y. Pi, *Phys. Rev. Lett.* 45: 1 (1980).

[15] S. Weinberg ref. [11]; R. N. Mohapatra and G. Senjanovic, *Phys. Rev. Lett.* 42: 1651 (1979); *Phys. Rev. D* 20: 3390 (1979).

[16] S.-Y. Pi, High Energy Physics—1980 (XX International Conference, Madison, Wisconsin), part 1, p. 505.

[17] S. Coleman, *Phys. Rev. D* 15: 2929 (1977).

[18] T. W. B. Kibble, *J. Phys. A* 9: 1387 (1976).

[19] A. H. Guth and E. J. Weinberg, ref. [12].

[20] P. Steinhardt, *Phys. Rev. D* 24: 842 (1981).

[21] S. Parke and S.-Y. Pi, *Phys. Lett. B* 107: 54 (1981).

[22] A. H. Guth, *Phys. Rev. D* 23: 347 (1981).

[23] A. H. Guth and E. J. Weinberg, ref. [12].

[24] A. D. Linde, *Phys. Lett.* 389 (1982).

[25] S.-Y. Pi, to be published in *Phys. Lett.*; A similar phase transition has been found in a $SO(3)$ supersymmetric model by P. Ginsparg, Harvard preprint HUTP-81/A053 (1981).

[26] E. Witten, *Phys. Lett. B* 105: 267 (1981).

[27] S. Coleman and E. J. Weinberg, *Phys. Rev. D* 7: 788 (1973).

From Gell-Mann–Low to Unification

Asim Yildiz

An evolutionary picture of unification from the Gell-Mann–Low equation to grand unified theories is given and some unification questions are discussed.

I. Toward Unification

Almost thirty years ago, during the hot summer of 1953, Francis Low and Murray Gell-Mann, in a relatively cool basement room[1] of the student center of the University of Illinois, were feverishly working on a question in quantum electrodynamics which had something to do with the dynamics of bare and dressed electrons and particularly with the coupling strength. The result of that summer's work, though it may have seemed unrelated at that time, was probably one of the first sparks that has ignited contemporary unification theories in physics. Indeed, the Gell-Mann–Low equation [1] paved the road to the renormalization-group equations and eventually to asymptotic freedom and the first naive estimate of a unification scale at the Planck mass [2] for the strong and electroweak forces. Francis Low's interests in physics grew incessantly over the years in many directions. He has recently expressed a wish to hear about current activities in unification, and I would like to reply to his request in this paper.

Although the idea of unification has occupied human minds from the time of the ancient Greeks onward—and most notably in the Maxwell and Einstein eras—recent years have witnessed a rebirth of extraordinarily intensive effort on the unification of all forces. Especially after the successful theoretical work on electroweak unification and experimental confirmation of the theory by neutral-current events, major efforts have been directed toward the unification of the strong and electroweak forces.

1. Other office mates of the cool dungeon were none other than David Pines and T. D. Lee.

These forces are known to be represented by the symmetries $SU(3)_{strong}$ and $[SU(2) \times U(1)]_{electroweak}$.

Symmetry arguments in unification require the symmetries to be embedded in larger groups that have a unique coupling constant. From this basic requirement, the first naive estimate of unification energy was determined. Using the renormalization group one can obtain the unification energy by requiring that the coupling constants g_3, g_2, and g_1 as functions of momentum Q or mass scale have equal values at a high Q. Thus the grand-unification mass scale was found to be 10^{17} GeV. It was later found that radiative corrections [3] reduce this value to $\approx 10^{-15}$ GeV. This, in turn, led to a series of novel experimental techniques (supported by phenomenological predictions) to measure the lifetime of a proton. Indeed a dimensional estimate for the proton lifetime can be given as

$$\tau = \frac{1}{g_0^2} \frac{M_X^4}{M_{Pr}^5} \approx 10^{32 \pm 1} \text{ yr}, \tag{1}$$

where g_0^2 is the unification coupling constant at 10^{15} GeV and M_X is the unification mass.

Since the grand-unification (unification of three forces) symmetry should at least contain $SU(3) \times [SU(2) \times U(1)]$ subsymmetries, the immediate analysis for the behavior of coupling constants of the respective subsymmetries would be via the renormalization group equations

$$\frac{1}{g_R^2(M)} = \frac{1}{g_{0i}^2(M')} + \frac{2C_1^{(i)}}{(4\pi)^2} \left(\frac{11}{3} \ln \frac{M}{M'} \right) - \frac{8}{3(4\pi)^2} C_1^{(i)} \ln \frac{M}{M'}, \tag{2}$$

which involve the standard Gell-Mann–Low β-function coefficient $11/3$ C_1. C_1 and C_2 are Casimir operators in the gauge (adjoint) and fermion representations of the particular symmetry i, defined by

$$C_1^{(i)} \delta_{cd} = f_{abc}^{(i)} f_{abd}^{(i)}, \quad C_2^{(i)} \delta_{ab} = \text{Tr}(T_a^{(i)} T_b^{(i)}). \tag{3}$$

Note that at a symmetry-breaking scale, a single coupling constant appearing in two places becomes two (or more) coupling constants with different evolution equations, or vice versa. Thus at the unification energy (10^{15} GeV) it is expected that $SU(3)$ and $SU(2) \times U(1)$ coupling constants separate, while at the 10^2 GeV mass scale $SU(2) \times U(1)$ will break and recombine into $U(1)_{em}$.

This naive picture was the first to demonstrate the dynamical evolution of grand unification. Using a temperature scale, one can describe the evolution of the symmetry of the universe in the same manner. This fact immediately invited the incorporation of cosmology into particle physics, and the discovery of relations between cosmology and particle physics has emerged as one of the most active areas in recent years.

Here it is pertinent to discuss some of the important and less well understood questions of grand unification in particle physics and cosmology.

Maladies of Unification

From the point of view of particle physics, grand unification is burdened with several problems that have been actively pursued in recent years.

(i) Mass Problem Probably the most tantalizing mystery in particle physics has been the quasi-random structure of the mass spectrum of the leptons and quarks. It is well known that fundamental principles alone cannot provide a physical mass unit. However, the relation of one mass value to another is a long-standing target for scientific explanation. This "mass problem" has been attacked recently with novel tools such as current algebra, quark dynamics (especially interquark potential), bag model, etc.

Additional input for the resolution of the mass problem is expected to come from the symmetry arguments of unification. For instance, lepton-quark symmetry is realized in unification symmetries. In the same line of unification considerations, the bare lepton-quark mass ratios can be determined in some of the proposed unification models. Fine tuning of the physical masses is left to the radiative corrections. Despite all efforts, none of the unification models have provided a satisfactory explanation of lepton and quark masses. Some hopes lie in the direction of supersymmetry entering into the unification picture, but so far this scheme has not produced anything realistic.

(ii) Gauge Hierarchy From the start unified models containing $SU(3) \times SU(2) \times U(1)$ as subgroups require two mass scales, at 10^{15} GeV and 10^2 GeV, to explain low-energy phenomena such as the quark-lepton mass spectrum. However, no satisfactory mechanism has yet been proposed to break symmetries at both of these mass scales in a natural manner. Suggestions have included logarithmic potentials and dynamical symmetry-breaking with simultaneous spontaneous symmetry-breaking, but the problem remains unsolved.

(iii) Flavor Question So far three families of quarks ($u, d; c, s; t, b$) and of leptons ($e^-, \nu_e; \mu^-, \nu_\mu; \tau^-, \nu_\tau$) have been observed, except for the t-quark. Because there are three observed lepton families and because of lepton-quark symmetry, the observation of the t-quark is much anticipated.

Besides this experimental question, the biggest theoretical question has been how to embed three families without imposing repetitions arbitrarily.

Naive embedding of three families in a large-rank symmetry produces disasters in other aspects of unification. For instance, in $SU(N)$ theories for large N asymptotic freedom is lost. Extended theories $G \times G$, or $G \times G \times G$ types where G can be one of the first generation symmetries $SU(5)$, E_6, $SO(10), \ldots$, can provide an answer to the question; but such a system requires an undesirable discrete symmetry, which implies at least a divided universe with an impenetrable wall [5].

(iv) Higgs Scalars The usefulness of Higgs bosons in symmetry breaking and in renormalization of gauge theories have gained them an important role in the cast of particles in almost every scenario. Higgs bosons are expected to be discovered in every experiment. Theorists with great ardor poured out many predictions for their masses, anywhere from a few GeV to hundreds and thousands of GeV. In the absence of experimental verification of these particles, grand-unification models started to look into alternative possibilities to replace Higgs particles. The technicolor idea was a response to this need. A Higgs-like structure is assumed to be born out of condensation of two or four quarks further accompanied by a dynamical symmetry breaking, thus nullifying the need for fundamental Higgs particles in the unified models. This, in turn, opened new questions in particle theories. Higgs scalars still remain to be observed.

(v) Vector Bosons Weak interactions [5] are believed to be mediated by vector bosons. So far these objects (W^+, W^-, Z) have not been observed. Most likely the Z-particle should be detected soon in $p\bar{p}$ or pp, or even in $e^+ e^-$ experiments by the $\mu^+ \mu^-$ final state. It is almost definite that such particles exist. But a paranoic possibility whispers, "What if they are not found?" This seems, of course, absurd. No scalar particles can do the job. The vector-boson question is an experimental question, not any longer a theoretical one.

(vi) Massive Light Neutrinos Experimental verifications of three light neutrinos (v_e, v_μ, v_τ) over several decades are also supported by the Helium-abundance arguments of cosmologists. Some theorists don't take cosmological estimates seriously, since by persuasive arguments the allowed number of families can be extended beyond three. Recent years witnessed experiments performed in the United States and abroad to convince ourselves that the light-neutrino family can have or may have masses in the eV range. KeV masses would violate observational limits on the mass of the universe since the neutrino number ($n_v + n_{\bar{v}}$) is expected to be of the same order of magnitude as the photon number (n_γ) of the universe.[2]

Putting aside technical and experimental questions, if one tries to give such a small mass to light neutrinos, some of the first-generation unified models fail the test. Unified models which violate $B - L$ (baryon minus lepton)- or L-number conservation can provide masses for light neutrinos. Using intermediate mass scales as in $SU(7)$, it is possible to arrange dynamical processes plausible for the light-neutrino mass generation. Entrance of Yukawa couplings and very large mass scales (compared to eV energies) make such mechanisms less believable.

II. Cosmology and Unification

The unification question has always found a close partnership in the historical evolution of the universe. There are several areas of cosmology which give substantial, if not very quantitative, arguments to help complete the unification scenario. Furthermore, the cosmos is a rich laboratory, with its stars, black holes, and other celestial sources, for the testing of many questions related to unification.

In recent years, cosmology found common grounds with particle physics in the investigation of unification-related topics. The most prominent ones will be briefly discussed.

Matter-Antimatter Difference of the Universe

Recent measurements of the background temperature of space give about 2.7°K. The number of photons per cm^3 is calculable from

$$n_\gamma \approx 20T^3\,;$$

for $T = 2.7°$K, $n_\gamma \approx 500$ per cm^3. Now baryon number can be estimated from the density of the universe:[3] $\rho \approx 10^{-30}$ gm cm^{-3}. The baryon

2. Note, however, that $n_{\nu/e} - n_{\bar\nu/e}$ is likely to be quite small-probably of the order of $10^{-10}n_\gamma$ [6].
3. Einstein and de Sitter [7] obtained the critical matter density of the universe, i.e., the value that yields a (nonstatic) flat solution without a cosmological constant. The result is $\rho_c = 3H^2/(8\pi G)$, where G is the Newtonian gravitational constant and H is Hubble's constant. They found that ρ_c was, within observational uncertainties, compatible with the observed density; thus, observation does not give the sign, even, of the curvature. More recent observations still are compatible with zero curvature, although the "visible" mass density (mass in stars, other radiating objects, etc.) seems to be about one order of magnitude short of the critical value. Without going into details, the value of H thus gives us an approximate value of the actual density; $H = 100$ km sec^{-1}/mpc gives $\rho_c \approx 2 \times 10^{-29}$ gm cm^{-3}, and the baryon density will correspond to a mass density of $\approx 10^{-30}$ gm cm^{-3}.

(proton) mass is $m_p \approx 10^{-24}$ gm. Hence, the number of baryons per cm^3 is $n_B \approx 10^{-6}$ cm^{-3}.

Assuming that the universe is matter-dominated and that antimatter $n_{\bar{B}}$ is less than matter n_B, one can show that the net baryon number is[4]

$$\frac{n_B - n_{\bar{B}}}{n_\gamma} \approx \frac{n_B}{n_\gamma} \approx 10^{-8} - 10^{-9}.$$

The first proposal to explain the net baryon number from the quantum-theory point of view is due to Andrei Sakharov[5] who stressed CP-non-conserving irreducible processes as the source of the matter-antimatter difference of the universe. More recent activities by many physicists have emphasized the role of grand unification in understanding these processes.

4. Net baryon number or density is expressed conveniently in terms of the dimensionless ratio $n_B/n_\gamma \approx (n_B - n_{\bar{B}})/n_\gamma$. (Expressing baryon number as a fraction of original baryon plus antibaryon number is unsatisfactory because the latter is temperature dependent.) Alfven and Klein [8] have shown under reasonable assumptions that the net baryon (B minus $\bar{B}$) number to photon ratio will be related to the entropy per baryon, σ, produced in B annihilation:

$$\frac{1}{\sigma} = \left[\frac{n_B - n_{\bar{B}}}{n_B + n_{\bar{B}}}\right]_{\text{before}} \approx \frac{(n_B - n_{\bar{B}})_{\text{after}}}{2n_\gamma} \approx \frac{n_B}{2n_\gamma},$$

since before the annihilation era baryons were relativistic and obeyed an equilibrium Boltzmann law; thus $n_B \approx n_{\bar{B}} \approx n_\gamma$.
Simple expansion leaves density ratios invariant; other processes since $\bar{B}$ annihilation don't change n_B/n_γ significantly. Thus today's n_B/n_γ value (10^{-6} cm^{-3}/500 cm^{-3}) gives approximately the value of $(n_B - n_{\bar{B}})/(n_B + n_{\bar{B}})$ that must have resulted from the era in which net baryon number appeared.
5. Sakharov [9] was the first who saw the interconnection between CP violation and the baryon asymmetry of the universe. The simple argument of this discovery is as follows:
(i) A reaction which results in $B - L$ (C invariance) is

$A + B \to C + D \ldots \Delta B = +a,$
$\bar{A} + \bar{B} \to \bar{C} + \bar{D} \ldots \Delta B = -a.$

Unless C nonconservation occurs, no net baryon number will be generated.
(ii) The inverse reaction (T invariance) is

$A + B \to C + D \cdots \Delta B = +a.$
$C + D \to A + B \cdots \Delta B = -a.$

Unless T nonconservation occurs, no net baryon number will be generated.
Thus, in order to suppress these cancellations C and T invariances necessarily are to be violated. But CPT is conserved. Let H be the Hamiltonian of the system, then $O = [CPT, H]$ $= [CP, H]T + CP[T, H]$; since $[T, H] \neq 0$, then $[CP, H]$ must be nonzero. Thus CP nonconservation must occur.
Recent years have witnessed a plethora of papers pursuing Sakharov's discovery in grand-unification models. These recent studies have revealed that the most pertinent information to GUT model building is the necessity of multi-Higgs multiplets. See note 6. For a complete list of recent publications see [10].

It is true that if the phenomena are incorporated into a unified model, then Baryon- and/or lepton-nonconserving processes must be an integral part of the model. Although this stands out as a necessary condition, only particular interactions of the unified theory, namely Higgs-particle decays to baryon and lepton pairs, form the desired irreversible processes:

$$H \rightleftarrows B + L. \tag{4}$$

However, to obtain n_B/n_γ one needs to go to a one-loop process where, again, a Higgs exchange between baryon and lepton models is needed. The process that provides net baryon generation must also have CP-violation simultaneously. The second law of thermodynamics implies that equilibrium scattering will not contribute to the net baryon number, while cosmological departure from equilibrium is associated with an absence of scattering. Hence one looks for decay processes that are in agreement with the second law of thermodynamics. Other decay mechanisms than Higgs decay fail to give an acceptable value for n_B/n_γ ($\approx 10^{-8}$). For instance, vector-meson decay with vector-meson exchange in a one-loop process does not yield CP-nonconservation, and baryon number production is canceled by the decay of the antiparticle. Vector-meson decays with Higgs exchange in a one-loop process contain CP-nonconservation but yield $n_B/n_\gamma \approx 10^{-8} \times \alpha/2\pi \approx 10^{-11}$, which is far too small.

Thus only decay mechanism that gives an acceptable value for n_B/n_γ is Higgs decay with Higgs exchange in the loop. In order to observe the required CP nonconservation these two Higgs scalars must come from at least two different Higgs scalar multiplets.[6] This rather important observation has some interesting implications in unified theories.

Monopole Abundance of the Universe
Over half a century ago, Dirac discussed magnetic monopoles [13]. Start-

---

6. Because of irreversibility (nonequilibrium) coupled with CP nonconservation, the leading workable scheme requires interference between the Higgs H_1 simple decay to a baryon and a lepton, a vertex with coupling constant g_1' ($H_1 \rightarrow B + L$, amplitude M_1), and the same Higgs H_1 decay to a baryon and a lepton (coupling constant g_1) with a second Higgs H_2 interchanged between B and L legs with coupling constants g_2 and g_2^+ respectively with amplitude M_2. CP nonconservation is expected to be seen in the 2 Im $M_1 M_2^*$ expression:

$$M_1 M_2^* \approx \operatorname{Tr} g_1' g_2'^\dagger g_1^\dagger g_2,$$

which is not generally real. For one Higgs multiplet $g_1 = g_2$ and $g_1' = g_2'$ then $M_1 M_2^*$ is real and there is no CP violation.

This situation was first noticed by two groups working independently: Ignatiev, Kuzmin, and Shaposhnikov [11], and Yildiz and Cox [12].

ing with these suggestions of Dirac monopoles, further elucidiation of monopole theories (most notably Schwinger's monopole theory) described a monopole of mass about 1 GeV. After the invention of gauge theories, it became obvious that the static solution of non-Abelian theories yields a new kind of monopole solution. These gauge monopoles are much heavier than Dirac monopole and are of mass

$$m_m = m_X/\alpha \approx 10^{16} \text{ GeV}$$

where m_X is the mass of the superheavy gauge bosons with mass about 10^{14-15} GeV. Embedding monopole-production mechanisms into unified theories does not present any problem, but the monopole density of the universe poses challenging questions.

Several authors [14] in recent years investigated the monopole question by using gauge theories, temperature effects, etc., to give a limit on the primordial monopole-to-baryon ratio in the galaxy,

$$n_m/n_B \approx 10^{-20},$$

and thus determining the galactic flux of monopoles to be

$$F \approx 10^{-3} \text{ m}^{-2} \text{ yr}^{-1}.$$

The high monopole density and their rather large mass suggest that either in cosmic rays or in matter monopoles should be detectable. The lack of experimental confirmation for these objects suggests the existence of an overlooked suppressing mechanism of monopole production in the evolution of the universe [15]. These mechanisms and the monopole question as a whole have not affected the symmetry questions of grand unification models seriously. Future investigations may find more serious effects or set up restrictions on unified-model building.

Helium Abundance

The helium abundance of the universe sets up certain restrictions on the flavor question. The abundance of ^{4}He, mostly, but also of D, ^{3}He, and ^{7}Li reflects several facets of the evolution of the universe. First, when the universe was three minutes old, ^{4}He was produced by hydrogen burning (overall, $4p \rightarrow [2p + 2n] + 2e^+ + 2\nu_e$, and $[2p + 2n] = {}^4$He.) Second, the observed density ^{4}He (and temperature $3°$K) supports the hypothesis that the universe began from a hot big bang. Last, a most important aspect of helium abundance is its implication for neutrino physics, especially regarding the question of neutrino species.

Helium abundance is a sensitive function of the expansion rate of the universe in the primordial nucleosynthesis period. The number of neutrino species affects the expansion rate because the total energy density in neutrinos is proportional to the number of distinct neutrino types; and the expansion rate is governed by the total energy density.

If the expansion rate was high, temperature was falling rapidly, and little helium could accumulate before the temperature was too low for synthesis to continue. If the expansion rate was low, much more helium would form.

Also relevant is the expansion rate relative to neutron β decay. If the expansion is too slow, neutrons decay before they cool to the formation temperatures and no helium is formed.

As was mentioned earlier, an analysis of helium abundance indicates that up to four different species of neutrinos can be accommodated in our universe.

The horizon problem [16] and various abundance questions have much more to do with thermodynamics, flatness of the universe, and other esoteric aspects of the unified-field Lagrangian than the symmetry questions. Although cosmological investigations can confirm aspects of unified theories, the burden mostly lies on those models that claim to be unified, which models may possess cosmological production mechanisms for baryons, monopoles, etc. However, a significant cosmological signature is the helium abundance, which appears to set an upper limit for the number of neutrino species.

III. The Fourth and the Ultimate Force

Beyond the yet-unanswerable questions of grand unification, the paramount question is the unification of all forces. The question seems to lack an answer at present mainly because of the incomplete resolution of the renormalization program of gravitational interactions on one hand, and symmetry questions on the other hand. The two questions, seemingly inseparable, have in recent years been given separate attention.

Renormalization of gravity, despite some advances (especially with supergravity) have experienced serious setbacks. There is no definite proof of renormalization of gravity in all orders. On the symmetry frontier, activities are focused to find a group structure with a unique coupling constant to include all forces. These ended up with convoluted arguments and the reappearance of some old problems, such as massless composite particles, etc. Activities in unification seem to have seasonal shifts from one subject to another. Meanwhile, new problems seem to be accumulating under the unification question. A few examples are technicolor theories (a response to invisible Higgs particles), supersymmetry; supergravity, the strong CP problem (a residue of the instanton or Euclidean solitons), etc. The future, but probably the not-too-far future, will be filled with surprises and new discoveries, especially from experimentalists.

Acknowledgments

I wish to thank Mark Claudson and Paul Cox for their discussions. I also thank Harvard Physics Department members and staff for their hospitality. This research was supported by the U.S. Department of Energy under contract ER-78-5-02-4999.

References

[1] M. Gell-Mann and F. Low, *Phys. Rev.* 95: 1300 (1954). Most later discussions of the renormalization group involve this paper or its generalizations.

[2] H. Georgi, H. Quinn and S. Weinberg, *Phys. Rev. Lett.* 33: 451 (1974).

[3] Several groups independently performed radiative correction calculations. See W. J. Marciano, *Phys. Rev. D* 20: 274 (1979), and T. Goldman and D. A. Ross, *Phys. Lett. B* 84: 208 (1979).

[4] I. Yu Kobsarev, L. V. Okun, and Ya. B. Zeldorich, *Phys. Lett. B* 50: 340 (1974).

[5] Architects of the electroweak unified theory are Sheldon L. Glashow, *Nucl. Phys.* 22: 579 (1961); S. Weinberg, *Phys. Rev. Lett.* 19: 1264 (1967); and A. Salam, *Proceeding of the Eighth Nobel Symposium on Elementary Particle Theory*, edited by N. Svartholm, Wiley, New York (1969).

[6] See D. V. Nanopoulos, D. Sutherland, and A. Yildiz, *Lett. al Nuovo Cim.* 28: 205 (1980).

[7] A. Einstein and W. de Sitter, *Proc. Nat. Acad. Sci.* 18: 213 (1932).

[8] H. Alfven and O. Klein, *Ark. Fys.* 23: 187 (1962).

[9] A. Sakharov, *Zh. ETF Pis. Red.* 5: 22 (1967) [*JETP Lett.* 5: 24 (1967).

[10] T. Yanagida and M. Yoshimura, TU/80/211.

[11] A. Yu. Ignatiev, V. A. Kuzmin, and M. E. Shaposhnikov, *Zh. ETF Pis. Red.* 30: 726 (1979).

[12] A. Yildiz and P. H. Cox, *Phys. Rev. D* 21: 906 (1980).

[13] P. A. M. Dirac, *Proc. Roy. Soc. A* 117: 610 (1928). One of the important questions, namely, "Can the Dirac monopole form a bound state with an electron?" was studied by Harish-Chandra, *Phys. Rev.* 74: 883 (1948), who found a negative result. Upon the discovery of gauge monopoles by G. 't Hooft, *Nucl. Phys. B* 79: 276 (1974), and A. M. Polyakov, *Zh. ETF Pis. Red.* 20, 430 (1974) [*JETP Lett.* 20: 194 (1974)], it was found by R. Jackiw and C. Rebbi, *Phys. Rev. D* 13: 3398 (1976), that gauge monopole can form a bound state with an electron. Calculations for the excited states of a gauge monopole with an electron has been carried out by P. H. Cox and A. Yildiz, *Phys. Rev. D* 18: 1211 (1978).

[14] An excellent article about monopoles in the present and early universe is T. W. B. Kibble, ICTP/81/82–14, who discussed monopole productions at the phase transition in the early universe and monopole production relation to grand unification.

[15] Suppression mechanisms of monopole production are proposed by several authors. Again, ref. [4] contains a very lucid examination of the suppression mechansim and relevant references.

[16] The horizon problem is investigated by several authors, most notably by Alan H. Guth, *Phys. Rev. D* 23: 347 (1981). For a review of the subject and further references, see ref. [4].

Preons and Supersymmetry

Jogesh C. Pati,
Abdus Salam, and
J. Strathdee

1. An important aspect of preonic theories is the construction of composite fields and the commutation relations [1] among them, using preonic fields (with their canonical commutation relations) as input. Superfields appear to be ideally suited for playing the role of preonic fields. The basis for this is the remarkable group property possessed by chiral superfields.

In this paper we shall assume that supersymmetry holds for preonic fields and that it is broken just below the ionization energy (possibly 10^5 GeV or higher) for the formation of quarks and leptons as preonic composites [2].

2. The preonic theory we choose to illustrate these remarks with is the theory of three preon types [2]: f (flavons), c (chromons) and the singlet s (henceforth called the "drone"). These are (2, 1, 1), (1, 2, 1), (1, 1, 4), and (1, 1, 1) representations of $SU_L(2) \times SU_R(2) \times SU_C(4)$. In addition each one of the preons carries $U(1) \times U(1)$ quantum numbers which permit of their binding into quark/lepton composites. The quarks and leptons themselves are neutral relative to these $U(1)$'s. The difficult problem in supersymmetry theories always is the breaking of this symmetry. By using these $U(1)$'s we shall show that it is possible to break supersymmetry as well as $SU(4)$ (and some of the $U(1)$'s) *simultaneously*. Thus if the scale of $SU_C(4)$ spontaneous breaking is of the order of 10^4–10^5 GeV [3], this could also be the scale of spontaneous breaking of supersymmetry, in contrast to other recent attempts [4] which break supersymmetry either at the $SU(3) \times SU(2) \times U(1)$ level of 300–1000 GeV, or assume that it is broken only at Planck energies.

3. Let Φ_- and Φ_+ represent left- and right-handed [5] superfields, each describing particles of spin $\frac{1}{2}$ and 0. The (Majorana) θ expansion of these fields is

$$\Phi_\mp (x, \theta) = \exp\left\{\pm \frac{1}{4}(\bar{\theta}\partial\gamma_5\theta)\left[A_\mp(x) + \bar{\theta}\psi_\mp(x) + \frac{1}{2}\bar{\theta}\left(\frac{1 \mp i\gamma_5}{2}\right)F_\mp(x)\right]\right\}.$$

The chiral fields are annihilated by covariant operators $D_\mp$, i.e.,

$$D_-\Phi_+ = D_+\Phi_- = 0,$$

where

$$D_\pm = \frac{1 \pm i\gamma_5}{2}\left(\frac{\partial}{\partial\bar{\theta}} - \frac{i}{2}\partial\theta\right).$$

Note that Φ_-^* behaves like Φ_+ so far as chirality is concerned. In particular the operation D_-D_- on Φ_- gives rise to a field of the plus type Φ'_+.

The crucial property for our purposes is the group property

$$\Phi_-(x, \theta)\Phi'_-(x, \theta) = \Phi''_-(x, \theta),$$

and likewise for the fields $\Phi_+(x, \theta)$. Thus any product field created by multiplying any number of $-$ (or $+$) type of preonic chiral fields leads to a $-$ (or $+$) type of composite. *Chiral spin-$\frac{1}{2}$ and spin-0 preons composed supersymmetrically in this manner do not give rise to any spin-1 composites.*

Spin-1 composites can of course be constructed by multiplying preonic superfields of opposite chiralities. Thus a product field of two preonic fields Φ_+ and Φ_- gives rise to a general superfield $\Phi(x, \theta)$ describing spin-0, spin-$\frac{1}{2}$ as well as spin-1 objects. Such general superfields can be decomposed through chiral plus "transverse vector" projections as follows.

Define

$$E_+ = -\frac{1}{4\partial^2}(D_-D_-)(D_+D_+),$$

$$E_- = -\frac{1}{4\partial^2}(D_+D_+)(D_-D_-),$$

$$E_1 = 1 - E_+ - E_-$$
$$(E_\pm^2 = E_\pm);$$

then

$$\Phi(x, \theta) = (E_- + E_+ + E_1)\Phi(x, \theta)$$
$$= \Phi_- + \Phi_+ + \Phi_1,$$

where

$$\Phi_1(x, \theta) = A_1(x) + \bar{\theta}\psi_1(x) + \frac{1}{4}\bar{\theta}i\gamma_v\gamma_5\theta A_{v1}(x)$$

$$+ \frac{1}{4}\bar{\theta}\theta\bar{\theta}(i\partial\psi_1) + \frac{1}{32}(\bar{\theta}\theta)^2\partial^2 A_1.$$

The field Φ_1 contains spin-0 $(A_1(x))$, spin-$\frac{1}{2}$ $(\psi_1(x)$, Majorana, $\psi = C^{-1}\bar{\psi}^T)$, and spin-1 $(A_{v1}(x))$ pieces.

4. When dealing with symmetry groups of the $SU_L(n) \times SU_R(n)$ variety, cancellation of anomalies often requires the introduction of additional mirror fields [6]. These are fields with the same transformation character under $SU(n)$ as the original fields but carrying opposite chirality. Thus given a $\Phi_-(x)$ preonic field, a composite mirror field Φ_+^M can be constructed by using the singlet drone field s_- in the following manner:

$$\Phi_+^M = D_- D_-(\Phi_- s_-).$$

This construction is of course, not unique; in fact all composites $D_- D_-(\Phi_- s_-^r)$ are possible mirror fields, as indeed are the fields $E_-(\Phi_- s_-^{*r})$. If the singlets s_- carry $U(1)$ quantum numbers for binding to Φ_-—as in practice they will (see later)—the composite mirror fields will be distinguished from each other through their $U(1)$ labels.

Another distinguishing label is provided by f, the **F** number [7] (associated with the operator $i\bar{\theta}\gamma_5(\partial/\partial\theta)$), which may be defined for a general superfield as a combination of an intrinsic gauge transformation $e^{if\alpha}$ combined with the transformation $\theta \to \exp\{-\alpha\gamma_5\}\theta$. Thus let

$$\Phi(x, \theta) \to \Phi'(x, \theta) = e^{if\alpha}\Phi(x, \exp\{\alpha\gamma_5\}\theta).$$

Expanding $\Phi(x, \theta)$ into components,

$$\Phi(x, \theta) = A(x) + \bar{\theta}\psi(x) - \frac{1}{4}\bar{\theta}\theta F(x) + \frac{1}{4}\bar{\theta}\gamma_5\theta G(x)$$
$$+ \frac{1}{4}\bar{\theta}i\gamma_v\gamma_5\theta V_v(x) + \frac{1}{4}\bar{\theta}\theta\bar{\theta}\chi(x) + \frac{1}{32}(\bar{\theta}\theta)^2 D(x),$$

we can read off **F** numbers associated with the components:

$$\mathbf{F} = f \quad \text{for} \quad A, \ V_v, \ D$$
$$= f + 1 \quad \text{for} \quad \psi_-, \ \chi_+$$
$$= f - 1 \quad \text{for} \quad \psi_+, \ \chi_-$$
$$= f + 2 \quad \text{for} \quad F + iG$$
$$= f - 2 \quad \text{for} \quad F - iG.$$

(Here $\psi_{\mp} = \frac{1}{2}(1 \mp i\gamma_5)\psi$ and likewise for $\chi_{\mp}$.) From this we read that for a chiral field $\Phi_-(x)$, carrying intrinsic **F** number f, the **F** numbers associated with the components A_-, ψ_-, and F_- are f, $f+1$, and $f+2$. For $\Phi_+(x)$ with intrinsic F number f, the components A_+, ψ_+, F_+ carry f, $f-1$, and $f-2$, respectively, while the covariant operators D_- and D_+ add F numbers -1 and $+1$ to the fields they act upon.

To take an example, assume Φ_- carries intrinsic F number f, while the singlet s_- carries f_s. Then the mirror composite defined as $D_-D_-(\Phi_-s_-)$ has intrinsic F number $f+f_s-2$ and the alternative mirror $E_-(\Phi_-s_-^*)$ carries $f-f_s$. The F number assignments differentiate the two types of fields.

All gauge Lagrangians conserve F number provided the gauge field carries intrinsic $F = 0$, with the component assignments A_v, $D = 0$. The projection χ_- of the Majorana gaugino must be assigned F number -1 for the conservation to hold. The renormalizable matter Lagrangians $(\Phi_-\Phi'_-)_F = D_-D_-(\Phi_-\Phi'_-)$ and $D_-D_-(\Phi_-\Phi'_-\Phi''_-)$ conserve F number provided (in an obvious notation) $f_- + f'_- - 2 = 0$ in the first and $f_- + f'_- + f''_- - 2 = 0$ in the second term. Likewise for matter Lagrangians $D_+D_+(\Phi_+\Phi'_+)$ and $D_+D_+(\Phi_+\Phi'_+\Phi''_+)$, the corresponding requirements are $f_+ + f'_+ + 2 = 0$ and $f_+ + f'_+ + f''_+ + 2 = 0$.

To summarize,[1] if gauge particles A_v, D carry $F = 0$, while the gauginos $\chi_\mp$ carry $F = \mp 1$ ($\chi_- = C\bar{\chi}_+^T$, Majorana condition), then F number is conserved for gauge Lagrangians. For matter fields Φ_- and Φ_+, assign intrinsic F numbers $f_\mp$. The components $A_\mp$, $\psi_\mp$, $F_\mp$ then carry $f_\mp$, $f_\mp \pm 1$, $f_\mp \pm 2$. For conservation of F number in pure-matter renormalizable interactions, e.g., $(\Phi_-\Phi'_-\Phi''_-)$ we need to satisfy conditions like $f_- + f'_- + f''_- - 2 = 0$.

5. To make quarks and leptons $[(2, 1, 4)_L$ and $(1, 2, \bar{4})_L]$ out of preons:[2]

---

1. The conventional parity operation does not commute with the F operation as defined. For Abelian or non-Abelian gauge Lagrangians containing matter multiplets of Φ_- as well as Φ_+ type (and with F-number assignments for matter and gauge fields as in the text) a parity operation can be defined for mixtures of scalar and spinor components, in a manner so as to conserve both parity and the F number. In [5] and [7] however, we preserved the commutativity of F number and the parity operation by working with $N = 2$ extended supersymmetry, where the gauge field Ψ is supplemented with a chiral multiplet s_+ (in the adjoint representation of the internal symmetry) such that the gauginos are 4-component Dirac—rather than Majorana—particles. In [7], $N = 2$ extended supersymmetry was called by us complex supersymmetry.

2. If one were economy-minded, the preons c'_- could themselves be considered composites: for example $E_-(c_-^* s_-)$ so far as the $SU(2) \times SU(2) \times SU(4)$ quantum numbers are concerned. In the appropriate range of energies where both c_- and c'_- are considered elementary and structureless, we shall of course have to ensure that the $U(1)$ and other quantum numbers of c_- and c'_- match; see section 6. For example, the f label of c'_- constructed as above (with $r = 1$) is $f_s - f_-$. This would equal the f label of c_- only if $f_- = \frac{1}{2}f_s$.

$$f_- = (2, 1, 1),$$
$$f'_- = (1, 2, 1),$$
$$c_- = (1, 1, 4),$$
$$c'_- = (1, 1, \bar{4}),$$
$$s_- = (1, 1, 1),$$

we need, for example, the composites $(f_- c_- s^r_-)$ and $(f'_- c'_- s^{r'}_-)$ where r and r' are arbitrary positive numbers, so far as $SU_L(2) \times SU_R(2) \times SU_C(4)$ transformations are concerned. We should additionally assign $U(1) \times U(1) \times \cdots$ quantum numbers to the f's, the c's, and the s's such that the attractive $U(1)$ forces bind them into quarks and leptons $(2, 1, 4)$ and $(1, 2, \bar{4})$, with the further proviso that the composite quarks and leptons are also $U(1)$ neutral. These particular requirements are easily met: what is difficult to achieve is the orderly (spontaneous) breaking of super-symmetry as well as of internal symmetries.

To show that this may be done in principle, we demonstrate that it is possible to assign the $U(1)$'s in such a manner that the four-color baryon-lepton symmetry $SU_C(4)$ and supersymmetry are broken *simultaneously* by the same set of Higgs particles, possibly at energies around 10^4–10^5 GeV. To show this we follow a procedure [8] devised by Weinberg to break $SU(3) \times SU(2) \times U(1)$ symmetry and supersymmetry together. Weinberg employs the Fayet-Iliopoulos mechanism [9], with an extra $\tilde{U}(1)$ together with two super-Higgs multiplets. The quarks and leptons remain massless, even though other particles acquire masses.

6. Henceforth we will ignore $SU_L(2) \times SU_R(2)$ and the flavons. The chromons are

$$c_- \approx 4_{0,x},$$
$$c'_- \approx \bar{4}_{0,x},$$

where $(0, x)$ specify $U_Y(1) \times U_{Y'}(1)$ quantum numbers. We take two Higgs multiplets with quantum numbers

$$H_- \approx 4_{1,1},$$
$$H'_- \approx 4_{-1,1},$$

and a singlet

$$s_- \approx 1_{0,-2}.$$

In this model electric charge is $Q = T_4^4 + Y$, which (we show) will remain unbroken. For $U_{Y'}(1)$ anomaly cancellations we need mirrors, but these will be ignored. The gauge Lagrangian plus the matter terms $h(H'_- H_- S_-)_F$ give rise to the following potential for the scalar components of the appropriate multiplets:

$$V = \frac{G^2}{2}\left[(\bar{H}H)^2 - 2|H'H|^2 + |H'\bar{H}'|^2 - \frac{1}{4}(\bar{H}H - H'\bar{H}')^2\right]$$
$$+ \frac{g^2}{2}[\bar{H}H - H'\bar{H}' - \xi']^2$$
$$+ \frac{g'^2}{2}[\bar{H}H + H'\bar{H}' - 2|S|^2 - \xi'']^2$$
$$+ h^2[(\bar{H}H + H'\bar{H}')|S|^2 + |H'H|^2].$$

Here G, g, and g' are $SU_C(4)$, $U_Y(1)$, and $U_{Y'}(1)$ gauge-coupling parameters. After some work, the potential minimizes, with $SU(4) \times U_Y(1) \times U_{Y'}(1)$ breaking down to $SU_C(3) \times U_Q(1)$ with

$$\langle H \rangle = \begin{bmatrix} 0 \\ 0 \\ 0 \\ a \end{bmatrix}, \quad \langle H' \rangle = \begin{bmatrix} 0 \\ 0 \\ 0 \\ a' \end{bmatrix}, \quad \langle S \rangle = 0,$$

where

$$0 = 2g'^2(|a|^2 + |a'|^2 - \xi'') + h^2(|a|^2 + |a'|^2)$$
$$0 = \tfrac{3}{4}G^2(|a|^2 - |a'|^2) + g^2(|a|^2 - |a'|^2 - \xi')$$
$$+ g'^2(|a|^2 + |a'|^2 - \xi'') + h^2|a'|^2.$$

The quadratic terms in V giving masses for the 3 and $\bar{3}$ scalar components of H and H' are

$$(G^2 - h^2)(a'\bar{H} - a H') \cdot (a'H - a\bar{H}').$$

These terms are positive definite provided we impose the requirement $G^2 > h^2$. The quadratic (mass) terms in V which mix $\text{Re}\,H_4$, $\text{Re}\,H'_4$, $S_1 = \text{Re}\,S$, $S_2 = \text{Im}\,S$ are

$$(\tfrac{3}{4}G^2 + 2g^2 + 2g'^2)(a(\text{Re}\,H)^2 + a'(\text{Re}\,H')^2)$$
$$- (\tfrac{3}{2}G^2 + 2g^2 - 2g'^2 - 2h^2)2aa'(\text{Re}\,H)(\text{Re}\,H')$$
$$+ 2h^2(a^2 + a'^2)(S_1^2 + S_2^2).$$

One can check that positivity is assured since

$(2g'^2 + h^2) > 0.$

The Goldstone fields Im H and Im H' are massless.

Finally the inclusion of the chromons $c_{0,x}$ and $c'_{0,x}$ gives for mass terms of their scalar components:

$$|c_4|^2 [\tfrac{3}{4}(a^2 - a'^2) + x(a^2 + a'^2 - \xi'')]$$
$$+ |\bar{c}'_4|^2 [-\tfrac{3}{4}(a^2 - a'^2) + x(a^2 + a'^2 - \xi'')]$$
$$+ |\mathbf{C}|^2 [-\tfrac{1}{4}(a^2 - a'^2) + x(a^2 + a'^2 - \xi'')]$$
$$+ |\bar{\mathbf{C}}'|^2 [\tfrac{1}{4}(a^2 - a'^2) + x(a^2 + a'^2 - \zeta'')].$$

Clearly x can be chosen such that the positivity of these terms is ensured. The spin-$\tfrac{1}{2}$ components of the chromons are desirably massless at this stage. There are Goldstinos, but we do not discuss them in this paper.

To summarize, we have demonstrated that it is possible to break supersymmetry as well as an internal symmetry (like $SU_C(4) \times U_Y(1) \times U_{Y'}(1)$ down to $SU_C(3) \times U_Q(1)$) with the same set of Higgs particles, while the spin-$\tfrac{1}{2}$ chromons remain massless. The lesson of the foregoing calculation lies for us in its stressing yet again the role of the $U(1)$'s. These appear as a necessary feature for simultaneous supersymmetry and internal symmetry breaking, in addition to being needed for providing forces to bind preons together. In a recent set of papers [10], it was suggested that such $U(1)$'s may possibly be associated with (analogue) electric and magnetic-monopole charges for preons. In a further paper we propose to consider the role of supersymmetry in the context of such a preonic theory.

Note. One may inquire how far back the notion of pre-pre- ... preons may be carried. Can we eventually arrive at a single "monotheistic" chiral supermultiplet s_- carrying none other than $U(1)$ charges? In such an approach gauge groups like $SU_L(2) \times SU_R(2) \times SU_C(4)$ (as well as the vector mesons associated with them) are considered as arising at successive composite levels, the renormalizable interaction at each level being the leading part of an effective interaction based on an expansion in powers of the radii of the relevant composites.

Clearly a $U(1)$ gauging of a *single* chiral multiplet s_- will give rise to nonrenormalizable anomalies; thus s_- would need to be supplemented with a mirror supermultiplet s_+. Alternatively, one may conceive of a nongauge Yukawa (renormalizable) coupling s_-^3 and attribute the next level of composites to binding by this force. This is not an attractive suggestion; however, if the possibility of binding exists, there may arise at the second level, the mirror composites needed for a further $U(1)$ gauging. These are

$$\Phi_{\mp}^{(p)} = E_{\mp}(s_-^p(\bar{s}_- s_-)^r) \approx E_{\mp}(s_-^{p+r}\bar{s}_-^r), \quad r > 0, \quad p \geq 0.^3$$

Disregarding any problems connected with the *nonlocality* of the chiral projection operators $E_{\mp}$, $\Phi_{\mp}^{(p)}$ provide us with a pair of mirror fields (for each value of the $U(1)$ charge p). An anomaly-free gauge theory with a $U(1)$ composite-gauge vector multiplet may now be motivated at this level.

At the next level, starting with *one* pair $\Phi_-^{(p)}$ and $\Phi_+^{(p)}$, new bound-state composites comprising *three* pairs of mirror fields (each carrying $U(1)$ charges of magnitude pp') may arise. These are

$$(\Phi_-^{(p)})^{p'+r'}\overline{(\Phi_-^{(p)})}^{r'} \text{ and its mirror } (\Phi_+^{(p)})^{p'+r'}\overline{(\Phi_+^{(p)})}^{r'} \tag{A}$$

and

$$E_{\pm}\left[(\Phi_-^{(p)})^{p'+r'}\overline{(\Phi_-^{(p)})}^{r'}\right], E_{\pm}\left[(\Phi_+^{(p)})^{p'+r'}\overline{(\Phi_+^{(p)})}^{r'}\right]. \tag{B}$$

Assuming that there is a mass degeneracy among the four particles comprised in set (B), there is the possibility that in addition to the $U(1)$ gauge of the level before, there also exist the composite gauges $SU_L(2) \times SU_R(2)$, operative below the dissociation energy of the composites of set (B). In fact, if supersymmetry were broken at this stage, such that the fermions and the bosons contained in these four supermultiplets of set (B) were not *supersymmetrically* degenerate, the nonsupersymmetric composite-gauge group could be as large as $SU_L(2) \times SU_R(2) \times SU_C(4)$. The gauge particles associated with spin-0 bosons—which we have called $SU_C(4)$—would necessarily be pure vectors (rather than axial vectors). In this approach the distinction between *color and flavor quantum numbers would be a consequence of supersymmetry breaking.* Quarks and leptons would now form as nonsupersymmetric $U(1)$ neutral composites of these preons at the next level. Such a model is of course different from the model considered earlier in this paper, in that the level at which supersymmetry breaks is even prior to the emergence of $SU_L(2) \times SU_R(2) \times SU_C(4)$.

3. One must remember that there is a limiting relationship implied in the construction of composite fields; for example, $\Phi^{(1)}(x)$ is defined through the spacelike limit $x_1 \rightarrow x_2 \rightarrow x_3 \rightarrow x(s(x_1)(\bar{s}(x_2)s(x_3))$. There is an arbitrariness in the taking of this limit, mathematically reflected in the order in which the x_1, x_2, x_3 approach each other and physically representing the distinction of whether or not the $(\bar{s}(x_2), s(x_3))$ composite forms first. In this paper these possibilities have not been exploited.

The labels r and r' may be construed as generation labels, as has been suggested by a number of authors. To illustrate, assume that we start with $p = 1\Phi_-$, Φ_+ pre-preons. The composites $E_+(\Phi_-^{p'+r'}\Phi_-^{r'})$ and $E_+(\Phi_+^{p'+r'}\Phi_+^{r'})$ are formed as a consequence of the interplay of $(p'+r')r'$ mutually attractive and

$$\frac{(p'+r')(p'+r'-1)}{2} + \frac{r'(r'-1)}{2}$$

repulsive $U(1)$-forces among the constituents of the composites. Very naively, we can say that these forces balance when

$$r' = \frac{p'(p'-1)}{2}.$$

Thus, given p', the number of distinct generations r' may be limited by the relations

$$r' \leq \frac{p'(p'-1)}{2}.$$

One may ask the question: why supersymmetry in the first place for the "monotheistic" supermultiplet s_-? Our motivation for this—or rather for the stage when one works with s_- and its mirror s_+—has been in the context of the (analogue) dual Abelian electric and magnetic $U_E(1) \times U_M(1)$ theory of [10]. Disregarding supersymmetry, in such a theory, the "natural" pair of pre-preons would appear to be an electrically charged object $(e, 0)$ and a dual magnetically charged object $(0, g)$, at the first level. The preons (from which quarks and leptons are made) could then be the composites (e, g) of these, created through the electromagnetic forces $U_E(1) \times U_M(1)$. Assuming that the pre-preons $(e, 0)$, $(0, g)$ are spin-zero objects, the preons (e, g) would carry the field spin $|eg/4\pi| = N/2$ (N integer).

Now whatever the value of N, the *neutral* preon-anti-preon composites (which make up quarks and leptons) cannot carry any except integer (or zero) spins. This is because a neutral composite made from (e, g) and $(-e, -g)$ can have no field spin. In order that quarks and leptons do manifest half-integral spin, the set of preons or pre-preons must contain objects carrying *both* integer (or zero) as well as half-integer spins. This appears to motivate supersymmetry at the basic preonic or pre-preonic level. The implementation of this idea will need a supersymmetrization of the dual electric and magnetic theory.

References

[1] R. Delbourgo, Abdus Salam, and J. Strathdee, *Phys. Lett.* 22: 680 (1966).

[2] J. C. Pati, Abdus Salam, and J. Strathdee, *Phys. Lett. B* 59: 265 (1975).

[3] J. C. Pati and Abdus Salam, *Phys. Rev. D* 10: 275 (1974); J. C. Pati, Abdus Salam, and J. Strathdee, *Nucl. Phys. B* 185: 445 (1981).

[4] S. Dimopoulos and H. Georgi, Harvard preprint HUTP-81/A022 (1981); N. Sakai, Tokohu Univ. preprint TU 81/225 (1981); S. Dimopoulos and S. Raby, Santa Barbara preprint (1981); M. Dine, W. Fischler, and M. Srednicki, Princeton preprint (1981); and J. Ellis, M. Gaillard, L. Maiani, and B. Zumino in *Unification of the Fundamental Particle Interactions*, edited by S. Ferrara, J. Ellis, and P. van Nieuwenhuizen, Plenum, New York and London (1980), p. 69.

[5] In this paper we follow the notation and conventions of the review by Abdus Salam and J. Strathdee, *Fortschr. Physik* 26: 57 (1978).

[6] J. C. Pati and Abdus Salam, in *Proceedings of the EPS Conference on High Energy Physics, Palermo, June 1975*, edited by A. Zichichi; J. C. Pati, Abdus Salam, and J. Strathdee, Maryland preprint (1981).

[7] Abdus Salam and J. Strathdee, *Nucl. Phys. B* 87: 85 (1975); *Nucl. Phys. B* 97: 293 (1975); R. Fayet, *Nucl. Phys. B* 90: 104 (1975). Fayet's R number is essentially our F number.) In the present paper we work with general values for intrinsic F number, $f_\pm$ for chiral fields $\Phi_\pm$. Thus F number need not be tied to the fermion number of the spin-1/2 components of Φ_- and Φ_+.

[8] S. Weinberg, HUTP-81/A047 (1981).

[9] P. Fayet and J. Iliopoulos, *Phys. Lett. B* 51: 461 (1974).

[10] J. C. Pati, Abdus Salam, and J. Strathdee, *Nucl. Phys. B* 185: 416 (1981) and references therein.

Contributors

Jeremy Bernstein
Department of Physics
Stevens Institute of Technology

Geoffrey F. Chew
Department of Physics
University of California,
Berkeley

D. Danckaert
Institute of Theoretical Physics
University of Leuven, Belgium

P. DeCausmaecker
Institute of Theoretical Physics
University of Leuven, Belgium

Carleton DeTar
Department of Physics
University of Utah

John F. Donoghue
Department of Physics and
Astronomy
University of Massachusetts,
Amherst

Sidney D. Drell
Stanford Linear Accelerator
Center

Mitchell J. Feigenbaum
Los Alamos National
Laboratory

Herman Feshbach
Department of Physics
MIT

Val L. Fitch
Department of Physics
Princeton University

R. Gastmans
Institute of Theoretical Physics
University of Leuven, Belgium

Murray Gell-Mann
Department of Physics
California Institute of
Technology

Marvin L. Goldberger
President
California Institute of
Technology

Alan H. Guth
Center for Theoretical Physics
MIT

Kerson Huang
Center for Theoretical Physics
MIT

Robert L. Jaffe
Center for Theoretical Physics
MIT

R. Jackiw
Center for Theoretical Physics
MIT

Kenneth A. Johnson
Center for Theoretical Physics
MIT

Jogesh C. Pati
Department of Physics
University of Maryland

Adrian Patrascioiu
Department of Physics
University of Arizona

So-Young Pi
Lyman Laboratory of Physics
Harvard University
and
Research Laboratory of
Mechanics
University of New Hampshire

Abdus Salam
International Centre for
Theoretical Physics, Trieste
and
Imperial College, London

Gino Segrè
Department of Physics
University of Pennsylvania

J. Strathdee
International Centre for
Theoretical Physics, Trieste

W. Troost
Institute of Theoretical Physics
University of Leuven, Belgium

Steven Weinberg
Department of Physics
Harvard University
and
Smithsonian Astrophysical
Observatory
and
Department of Physics
University of Texas at Austin

William I. Weisberger
Institute for Theoretical Physics
State University of New York
at Stony Brook

Victor F. Weisskopf
Center for Theoretical Physics
MIT

Tai Tsun Wu
Department of Physics
Harvard University

Asim Yildiz
Lyman Laboratory of Physics
Harvard University
and
Research Laboratory of
Electronics
University of New Hampshire